DOUGLAS
DC-3 DAKOTA

1935 onwards (all marks)

ZENITH PRESS

Acknowledgements

Gathering the material for this publication would have been very difficult without the invaluable help and enthusiastic support of the following people: Tim Badham (Air Atlantique); Alf Bone, for his poignant insight into flying the Dakota in a combat scenario during the Second World War; Lincolnshire Aviation Heritage Centre, and Caroline Woodley (Aces High) for photographs and history of its aircraft; Trevor Morson, for his recollections of 'The Last Time'; Dave Peet, for his impressions of flying ZA947 prior to its arrival at the BBMF; Squadron Leader Stu Reid, former BBMF Dakota and Lancaster pilot, for his insight into flying the aircraft, and the crew's view; Andy Renwick at the RAF Museum for his help sourcing photographs; and Jarrod Cotter, Jim Douthwaite, Ian Hopwell, Dennis Lovesday and Richard Paver for their photographs. Our thanks to you all, and to our editor Jonathan Falconer for his patience and support during the process of writing this book.

Published in North America in 2011 by Zenith Press, an imprint of MBI Publishing Company, 400 1st Avenue North, Suite 300, Minneapolis, MN 55401 USA, by arrangement with Haynes Publishing.

Zenith Press titles are also available at discounts in bulk quantity for industrial or sales-promotional use. For details write to Special Sales Manager at MBI Publishing Company, 400 1st Avenue North, Minneapolis, MN 55401 USA.

To find out more about our books, join us online at www.zenithpress.com or www.qbookshop.com.

ISBN-13: 978-0-7603-4291-6

While every effort is taken to ensure the accuracy of the information given in this book, no liability can be accepted by the author or publishers for any loss, damage or injury caused by errors in, or omissions from, the information given.

Printed in the USA by Odcombe Press LP, 1299 Bridgestone Parkway, La Vergne, TN 37086

ZENITH PRESS

DOUGLAS
DC-3 DAKOTA

1935 onwards (all marks)

Haynes

Owners' Workshop Manual

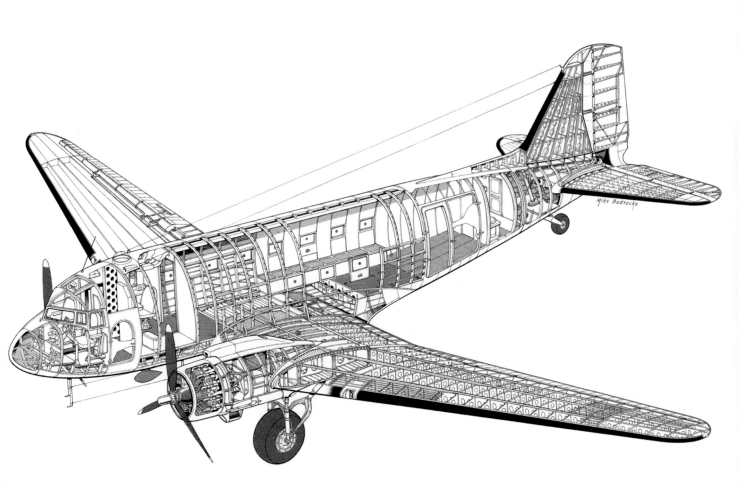

Mike Badrocke

An insight into owning, flying, and maintaining
the revolutionary American transport aircraft

Paul and Louise Blackah

Contents

OPPOSITE The rugged contours of Wings Ventures'
DC-3 D-Day veteran, N1944A, are seen to advantage in
this photograph. *(PRM Aviation)*

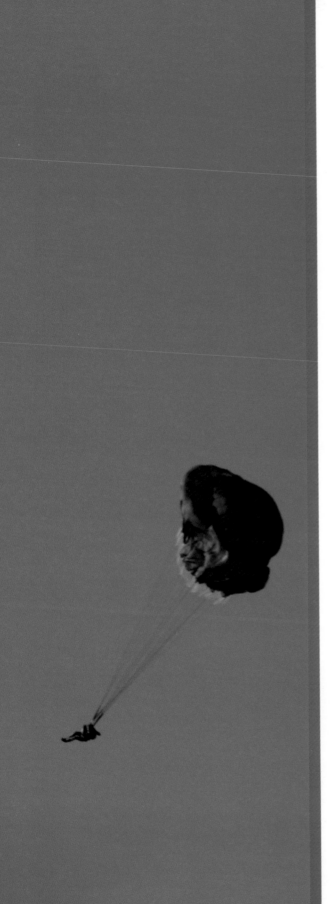

Introduction

Looking back at the significant aircraft of the Second World War, most would cite the fighters and bombers as the aircraft that won the war. It was, however, equally vital to have reliable aircraft for the transportation of men and equipment. The Douglas DC-3 Dakota filled this role. Its reliability and load-carrying capability made it an invaluable tool of Allied air forces in the Second World War. To this day the Dakota has a worldwide fan base and is still in use in many areas across the globe, fulfilling a variety of different practical and commemorative roles.

OPPOSITE **Para-dropping over Arnhem, 2010.**
(BBMF/Crown Copyright)

The DC-3 Dakota has enjoyed a career lasting for more than half a century, which is a testament to its versatility in both civilian and military service. Individual people think of different events and roles that reflect this versatility when they recall the Dakota and the part it has played in conflict: para-dropping in both the Second World War and Vietnam War, for instance. The less glamorous aspects of the Dakota's service have been its ability to carry cargo, personnel – both able-bodied soldiers and the wounded – and its modification to tow, drop and recover gliders. US General Dwight D. Eisenhower is alleged to have said that four things won the Second World War: 'the bazooka, the Jeep, the atom bomb and the C-47 "Gooney Bird".'

For the sake of clarity, throughout this book we will refer to the aircraft as a DC-3 Dakota. The Douglas Company named its aircraft the Douglas Commercial (DC-1, 2, 3) as the design was initially intended for commercial use. When the aircraft came into service with the military, each variant was prefixed by the letter 'C' (hence C-47 in US military use) and, because of the large number of designs there were many C-variants. In the United Kingdom the aircraft was known as the DC-3 Dakota, and as we are focusing on the United Kingdom's operation of the aircraft, specifically during the Second World War, we will continue with this tradition.

DC-3 Dakota ZA947

ZA947 has had a varied life, culminating in her role with the Battle of Britain Memorial Flight (BBMF). She was manufactured in March 1942 and, although initially issued to the United

States Army Air Force (USAAF), she soon became the property of the Royal Canadian Air Force (RCAF), operating with the serial number '661'. She served in Canada and then Europe until purchased by the Royal Aircraft Establishment (RAE) at Farnborough in 1971.

The RAE gave her the serial number KG661, but there was some doubt about the provenance of this serial. Research was undertaken that showed the original Dakota KG661 had been destroyed in an accident during take-off from Crosby-on-Eden,

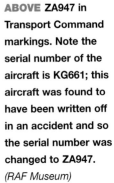

ABOVE ZA947 in Transport Command markings. Note the serial number of the aircraft is KG661; this aircraft was found to have been written off in an accident and so the serial number was changed to ZA947. *(RAF Museum)*

LEFT ZA947 in the Royal Aircraft Establishment colours. She was also called the 'Portpatrick Princess' at this time. *(RAF Museum)*

Cumberland, in December 1944, so in 1979 the new serial number ZA947 was allocated. During her time with the RAE, ZA947 was used for a variety of purposes, including para-dropping, trials, launching remotely piloted vehicles, and in static displays.

In 1992 the Defence Research Agency (DERA, the successor to the RAE) offered the aircraft for disposal and she became the property of RAF Strike Command, which officially handed the aircraft to the BBMF in March 1993. Since then ZA947 has become an important part of the flight, fulfilling many roles, not least that of displaying and commemorating within her own right. She is also used to perform general support tasks, such as ferrying ground crew and equipment to air shows, and playing a pivotal role in training aircrew for the Lancaster. This ensures that pilots of the bomber keep their hours 'current' when the Lancaster is unavailable for training purposes.

Every year ZA947 travels to Holland to take part in the annual commemoration of Operation 'Market Garden' at Arnhem, where she drops parachutists who are often dressed in period uniform. In 1993 she wore the markings of 271 Squadron's YS-DM, KG374, which was originally flown by Flt Lt David Lord VC, DFC. Lord was awarded a posthumous VC for bravery under fire on 19 September 1944. ZA947 has also taken part in many of the Berlin Airlift celebrations over the years.

In 2004 ZA947's interior was refitted

ABOVE ZA947 being given her correct serial number in 1979. As you can see, she is still in Transport Command markings. *(RAF Museum)*

LEFT ZA947 at Eindhoven during the 2009 Arnhem commemorations. *(Paul Blackah/Crown Copyright)*

OPPOSITE ZA947 carrying out one of the many commemorative fly-pasts to remember those who gave their lives during the Second World War. *(BBMF/Crown Copyright)*

LEFT ZA947 in the markings of YS-DM, of 271 Squadron, the aircraft of Flt Lt David Lord VC. *(BBMF/Crown Copyright)*

BELOW The RAF Falcons freefall display team, using ZA947 as a jump ship. *(BBMF/ Crown Copyright)*

with original and authentic para-seats, thus restoring her to a wartime specification. In 2008, after successfully dropping the RAF Falcons Parachute Display Team, permission was given for her to use the latest military static line parachute, which now enables the BBMF Dakota to continue her important role of commemorating those acts of bravery and commitment undertaken by veterans of the Second World War.

New markings

During the winter season of 2010/11 ZA947 left the BBMF for a stay at Humberside Airport with Eastern Airways, where she underwent a major service. After this in-depth service in which the aircraft is stripped back to the bare metal, the BBMF aircraft are repainted in new markings, representing a squadron and an aircraft that saw service in the Second World War.

For the next ten years ZA947 will represent FZ692, *Kwicherbichen*, of 233 Squadron. The squadron was in RAF Transport Command and played many roles during the Second World War. FZ692 herself served with the squadron from February until September 1944 when she was transferred to 437 Squadron RCAF.

FZ692 and the rest of 233 Squadron played a major role in the D-Day landings on 6 June 1944, towing gliders and dropping paratroops of the 3rd British Parachute Brigade into Normandy. As the day progressed the squadron flew a further 21 sorties to air-drop supplies to the British 6th Airborne Division. Four aircraft and crews were lost on this day alone.

As the battle progressed, ground was gained

and it became possible to land the aircraft in France. FZ692 and the rest of the squadron were able to resupply the troops on the ground and act as air ambulances on the return flights, evacuating the wounded, which once more demonstrated the flexibility of the Dakota and its design. The BBMF's Dakota wears her new colours with pride and in recognition of those who served and did not return.

It is interesting to note that the original FZ692 survived the Second World War and is still flying under the civilian registration, C-GRSB, and at the time of writing is owned by Environment Canada.

TOP RIGHT The original FZ692 *Kwicherbichen* of 233 Squadron. *(RAF Museum)*

RIGHT Close up of ZA947's new nose art. The red-and-white circles denote casualty evacuation trips, the bombs represent ammunition deliveries, the two parachutes denote supply drops, and the single crate is a supply delivery as opposed to a drop. *(Paul Blackah/Crown Copyright)*

BELOW ZA947 in her 2011 paint scheme as FZ692 *Kwicherbichen*. *(Jarrod Cotter)*

Chapter One

The Dakota story

The DC-3 Dakota may never have been built had it not been for the unswerving belief of Donald Douglas in his design for the aircraft. By the beginning of the Second World War the civilian version of the DC-3 had been hailed a success by commercial airlines. During the war the RAF considered acquiring other transport aircraft types, but its final choice of the Dakota proved to be a wise decision. Remembered particularly for its use by the USAAF and RAF on D-Day and at Arnhem, the DC-3 saw service worldwide during the Second World War.

OPPOSITE RAF Dakota Mk IIIs of 575 Squadron, based at Broadwell in Oxfordshire. During the early hours of D-Day the squadron sent 21 Dakotas on the initial drop of airborne troops and followed later that day with 19 glider-towing sorties. On 17 June, 575 Squadron made the first of its many casualty evacuation flights from Normandy back to Broadwell. *(RAF Museum)*

Donald Wills Douglas was born in Brooklyn, New York, on 6 April 1892. After graduating in 1909, Douglas enrolled in the Naval Academy at Annapolis, Maryland, and his future seemed set, but three years later he resigned to study aeronautical engineering at the Massachusetts Institute of Technology. He qualified in 1914.

Douglas's first job was with the Connecticut Aircraft Company, working on the design of the first dirigible, the DN-1. In August 1915 he left the company to take up a post as chief engineer at Glenn Martin Company. When the company merged with the Wright Company to form Wright-Martin he left to become the chief civilian aeronautical engineer of the Aviation Section of the US Army Signal Corps. His stay was brief and he soon left and moved to Cleveland, Ohio, to rejoin the newly re-formed Glenn L. Martin Company. Once again his stay was short and independence and a free hand beckoned.

In March 1920 Douglas took the courageous step of moving his family to southern California. He had less than $1,000, a firm belief in his engineering and design skills, together with the determination to find an investor in order to start his own company. An affluent gentleman named David R. Davis was introduced to Douglas and was impressed with his ambition and designs, and so the Davis-Douglas alliance began. With an investor and the expectation of orders for new aircraft, Donald brought six of

his former associates to California and his team was complete.

After initially working solely for the American armed forces, providing torpedo and observation aircraft, the Douglas Aircraft Company bid for a new passenger aircraft, which was ordered by Transcontinental and Western Airlines.

Donald's chief engineer was Arthur E. Raymond, with whom he had an excellent business relationship. Working together, it wasn't long before the first DC-1 was produced. The aircraft was fitted with the most up-to-date equipment of the time: new gyroscopic instruments, as well as the new Sperry automatic pilot. The fuselage measured 60ft from nose to tail, with a wingspan of 85ft. Inside the aircraft were two rows of six seats with a broad aisle between them. With a small galley and the novelty of a toilet, the DC-1 appeared the height of modern technology.

The two Wright Cyclone 710hp 9-cylinder, air-cooled radial engines powered the two Hamilton three-bladed propellers. Despite the aircraft's sleek lines and modern equipment, test pilot Carl A. Cover experienced major problems on the DC-1's first flight. Every time the nose dipped, the engine cut out and only his expertise saved both him and the aircraft. After detailed checks by the engineers it was found that the experimental carburettors were at fault; however, after turning their installation through 180 degrees the problem was permanently rectified.

Rigorous flight tests were carried out, including attempting take-offs and long-distance flights on one engine, and the DC-1 passed them all. TWA was so impressed with the aircraft that it took the DC-1 itself and ordered another 25, but they wanted certain modifications to be made. In order to carry 14 passengers instead of 12 they required a longer fuselage, which in turn necessitated the wings to be made 10ft longer. This new design was named the DC-2. The original DC-1 continued to fly until 1940, during which time she was sold on to new owners numerous times, until she crashed on take-off at Malaga and was written off.

TWA took delivery of the first DC-2 on 14 May 1934. So impressed were other airlines that the orders began to flood in. Not only

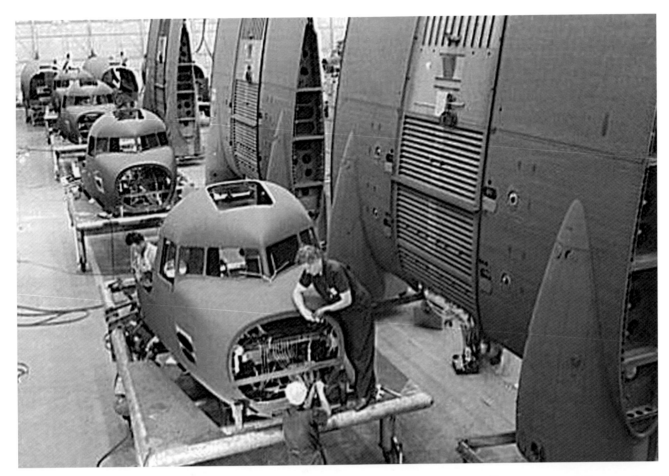

was the DC-2 used as a commercial passenger aircraft, but it was also entered into various air races. One of the more notable races was the 1934 MacRobertson Trophy Race, the London to Melbourne Derby. KLM, the Royal Dutch airline, was delighted when its DC-2 came in second after an easy run, demonstrating the aircraft's versatility.

As the volume of passengers increased, the airlines began to look for an aircraft that could carry more people. With a list of specifications devised by William Littlewood, chief engineer for American Airlines, the Douglas engineers set to modifying the DC-2 and the result was the DC-3, which first saw service for American Airlines on its Chicago to New York run in 1936. The aircraft proved to be a popular choice with many airlines and the Douglas Company now began to see some profit from its endeavours as orders poured in.

Across Europe the peaceful climate was changing and the first indications of possible conflict were beginning to emerge. Both the Americans and the British began assessing their military power and, realising that any conflict would require better-equipped air arms, they began to source possible aircraft to add to their respective air forces.

Modified to order

The DC-3 was a prime contender. Ease of modification, according to purpose, was a significant factor in the choice of the Dakota by America and Britain. France and Belgium also responded to Hitler's threat and placed orders with the Douglas Company, although before these contracts could be fulfilled the two nations had already fallen to the Nazis.

The British Purchasing Commission in the United States agreed to take on France and Belgium's contracts and also placed orders in its own right. By now it had been recognised that any future conflict would be far different from the First World War of 1914–18, and that aircraft would play a definitive role. An aircraft that could serve as a troop and cargo carrier would be vital.

ABOVE Nose sections and centre sections for C-47 transport planes on the production line at Long Beach. *(The Library of Congress LC-USE6-D-008115)*

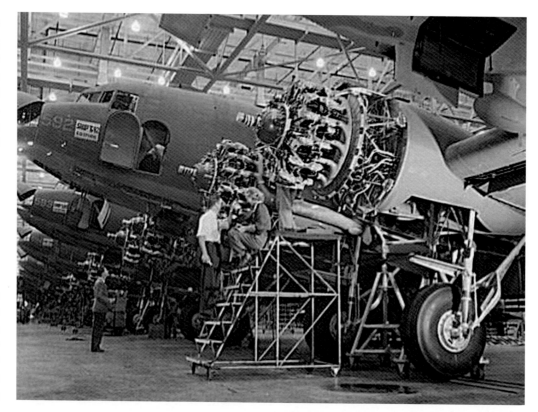

RIGHT C-47 production at Long Beach. Note the close spacing between each aircraft on the assembly line. *(Library of Congress LC-USE6-D-008124)*

BELOW A long line of C-47 transport planes awaiting the installation of their wings. *(Library of Congress LC-USE6 –D-008127)*

On 11 March 1941 the Lend-Lease Act was passed, which enabled the RAF and other Allies to acquire many more aircraft, including other types such as the Boeing Flying Fortress. Although as yet America was not playing an active role in the war, the Act was called an 'Act Further to Promote the Defence of the United States of America'. Hitler took this Act as a defiant stance from the United States and from then on his submarines were given free rein to attack US vessels.

A total of 2,000 Dakotas were delivered to the RAF under this scheme. At the same time Britain also purchased surplus Dakotas from the American civil aircraft operators, which were dismantled and shipped to the Middle East. Here – and also across India – they were used

alongside other DC-3s that were already in situ and being operated by Pan American Airways.

The DC-2 and DC-3 were modified according to their roles. These modifications are too numerous to mention here, but brief examples of some of them are listed below.

C-47 Basic military version of the DC-3 with seating for 27 troops; 965 were built, including 12 aircraft classified as R4D-1 that were put into service with the United States Navy (USN).

C-47A This was the basic C-47 model, fitted with a 24-volt electrical system; 5,254 were built, again with the USN designating their aircraft as R4D-5.

RC-47A C-47 basic model, but fitted with photographic reconnaissance equipment.

SC-47A C-47 basic model equipped for search-and-rescue, with an extra fuel capacity to cover the China–Burma–India routes.

C-47B Fitted with more powerful R-1830-90s engines with superchargers, and carrying extra fuel to cover the China–Burma–India routes.

C-47E Modified to carry 27–28 passengers or 18–24 litters.

The Dakota's role during the Second World War

The Dakota was used across the world in its many versions, and in a variety of front-line and support functions. Here, we will take a brief look at some of the better-known roles that the Dakota has played. It should be remembered that its support was crucial in many of the operations that determined the outcome of the Second World War. A full history of the military Douglas DC-3 Dakota would be a hefty tome if it included every combat operation that it undertook during the war, and in subsequent conflicts. Since we are primarily concerned with the workings and maintenance of the Dakota, the following

RIGHT The Dakota was used in many theatres during the Second World War, including the Middle East. *(Jonathan Falconer)*

accounts give a small taste of the aircraft's performance under fire.

Operation 'Overlord'

On 6 June 1944 the Allies launched Operation 'Overlord', the invasion of German-occupied France. The operation began in the early hours of the 6th with airborne assaults by the US 82nd and 101st Airborne Divisions at Ste Mère Église and the British 6th Airborne Division at Pegasus Bridge. There followed the largest amphibious assault in history along a 50-mile stretch of Normandy coastline in which almost 160,000 Allied troops were landed on five invasion beaches. The Dakota proved invaluable as paratroop transport and glider tug, as well as for dropping supplies and ammunition.

When the beachhead had been secured and the Allies had consolidated their position on French soil, they began the breakout. Once more the Dakota provided support as a personnel transport, supply carrier and

ABOVE Engine maintenance is carried out on an RAF Dakota by Thomas Lee, somewhere in the Middle East. *(RAF Museum)*

LEFT One of the main roles of the DC-3 was dropping paratroopers, for which it was used in particular on D-Day on 6 June 1944, at Arnhem on 17 September 1944, and later on the Rhine crossing in March 1945. *(Jonathan Falconer)*

RIGHT Grim-faced paratroopers of the US 101st Airborne Division – the 'Screaming Eagles' – pictured on their way to Normandy in a C-47 on the eve of D-Day. *(US National Archives)*

BELOW USAAF C-47s return at low level over the Normandy beaches to England, their mission accomplished. A fighter escort of Spitfires can be seen just behind the second Dakota. *(US National Archives)*

medevac. All Allied aircraft at this time had 'invasion stripes' (two black and three white) painted on the fuselage and wings for ease of recognition. The BBMF Dakota currently wears her invasion stripes with pride.

Operation 'Market Garden'

Operation 'Market Garden' (17–25 September 1944) was a carefully planned Allied airborne assault that aimed to capture key river bridges in enemy-occupied Holland. Its objective was to bring an early end to the war by enabling Allied troops to outflank the Siegfried Line and encircle Germany's industrial heartland of the Ruhr.

The D-Day landings in June 1944 had proved successful and Allied troops had broken out of the Normandy bridgehead by late August. The push into Germany was the next step. The primary aim of 'Market Garden' was the capture of vital bridges over the Meuse, the

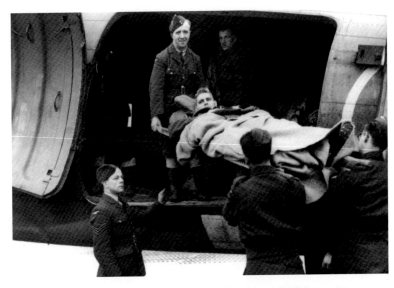

ABOVE Removing injured personnel from a Dakota at RAF Down Ampney. *(RAF Museum)*

BELOW Inside a Dakota on their way to Arnhem, British paratroops of the British 1st Airborne Division give the 'V'-sign and 'thumbs-up'. *(IWM K 7570)*

ABOVE American C-47s fly over St-Amand church at Gheel in Belgium on their way to Holland. *(IWM BU 918)*

Waal and the Lower Rhine, as well as bridges over several small tributaries and canals. To achieve this it was imperative to move vast amounts of troops and supplies into the area, but because these targets were inland behind enemy lines the only way in was by air.

It was decided to use gliders to carry personnel and supplies, as well as to drop

paratroops. Once again, this is where the versatility of the Dakota proved its worth. A custom-built towing attachment had been developed for the Dakota to enable it to pull a glider full of troops or equipment, and the pilots of both the Dakota squadrons and the glider units had been undergoing regular training together.

RIGHT This burnt-out Dakota Mk III, KG401, of 48 Squadron from Down Ampney, Gloucestershire, crash-landed in a field near Kessel, Holland, after parachuting supplies to beleaguered Allied troops at Arnhem on 19 September. The aircraft had just dropped its supplies from 700ft when it was brought down by intense anti-aircraft fire. *(IWM CE 165)*

On 17 September the combined airborne forces of the USAAF and RAF took off from airfields in southern England to begin the ill-fated mission. Together, they totalled 1,438 Dakotas. Some 160 Waco CG-4 Hadrian and 916 Airspeed Horsa gliders were towed by Dakotas and landed 14,589 troops, 1,736 vehicles and 263 items of artillery and heavy weaponry in the first wave. A further 20,011 troops parachuted into the area from Dakotas.

The first hours of the airborne assault were encouraging, with several bridges between Arnhem and Eindhoven captured, but then a combination of factors conspired to derail the whole operation. Overstretched supply lines, the demolition of a bridge over the Wilhelmina Canal, much stronger German resistance at Arnhem than was expected, and the failure of Allied ground forces to relieve the British 1st Airborne Division, saw their tenuous hold on the Arnhem bridgehead overrun and captured on 21 September. The rest of the division was trapped in a small pocket west of the bridge and was evacuated under cover of darkness on 25 September. 'Market Garden' was a heroic failure.

A veteran's view – Alf Bone

Alf Bone, a Second World War veteran pilot, flew the Dakota on a variety of operations, including 'Market Garden'. He recalls his memories of flying during that time.

ABOVE A Dakota about to pick up the tow rope for a WACO glider. *(RAF Museum)*

BELOW Veteran Dakota pilot Alf Bone in front of ZA947. *(BBMF/ Crown Copyright)*

I flew the Dakota as Captain from January 1944 to December 1945 on 271, 194 and 10 (India) Squadrons.

Quirks and foibles? None – almost faultless and a real lady. Combat scenario? Mostly I think of 'Market Garden', the Arnhem fiasco. On 17 September 1944 I tugged a Horsa glider from 271's base in Down Ampney, on 18 September 1944 another Horsa and on 20 September 1944 a resupply mission. The first two trips were pretty uneventful, but the third was dicey. We hadn't been told the drop zones were already in German hands, so that we were dropping supplies virtually down the muzzles of their 88mm flak guns. The squadron took a bashing and I lost two crews with whom I was very friendly. Flt Lt David Lord, also a comrade, was awarded a posthumous VC on one such resupply mission. At the end of the week many of our Daks were damaged, some destroyed, but I was unscathed. Perhaps because I flew lower than the briefed height, and though we saw a lot of tracers going up just past the wing, all we experienced was a hairy hailstorm of spent shrapnel bouncing off the metal roof of the aircraft.

I was in India taking out a Dak to support the Chindits, but didn't stay there. I therefore missed D-Day +10, taking over supplies and bringing back casualties. I have logged about 800 stretcher cases flown back from France, Belgium, Holland and eventually Germany. On VE-Day I was taking a load of carbolic soap to Celle, near Belsen. We heard Churchill on our radio announcing the end of the war in Europe as we flew over the Rhine. Two to three months later, with a different crew, I heard Atlee saying that the Japanese had capitulated as we flew over the Irrawaddy.

My time in Burma was short-lived, but hectic;

flying in impossible monsoon conditions. After that I volunteered to ferry troops from Madras to Kandi on their first leg of demobilisation with 10 Squadron, who had recently converted from Halifaxes on Bomber Command to Daks.

I therefore qualified for the France and Germany Star, the Burma Star and the medal I am most proud of, the Atlantic Star. During 1942–43 I flew Whitleys and Wellingtons on Coastal Command in pursuit of U-boats. The Whitleys were heavy to fly, the Wellingtons were marvellous aircraft, but not without vices, but the Dak was superb.

Operation 'Varsity'

Towards the end of the Second World War the Dakota took part in numerous operations, proving invaluable in both the transportation and supply of Allied forces. When it became evident that German forces, positioned at strategic points on the River Rhine, would seriously hamper the Allied advance, Operation 'Plunder' was put into place.

'Plunder' was the overall codename given to the task of crossing the Rhine and forging ahead into northern Germany, under the leadership of Field Marshal Montgomery. It began on 23 March 1945 and was followed a day later by Operation 'Varsity', the largest airborne operation in history to be carried out at a single location on a specific day. Several thousand aircraft dropped 16,000 paratroops and their equipment either by parachute or glider.

The British 21st Army Group was supported by the British 6th Airborne Division, which included a Canadian Parachute Battalion, and the US 17th Airborne Division. The paratroops were dropped simultaneously behind the German lines, close to the towns of Hamminkeln and Wesel. By the time the drop had begun the German divisions on the ground were markedly depleted and morale was low, but they did possess effective anti-aircraft guns, which were a big factor in the heavy losses among the Allies: 1,111 dead, 1,625 wounded and 50 aircraft and gliders shot down.

Once the Allied forces had established a bridgehead on the opposite bank of the Rhine, they advanced north into Germany, resupplied by Dakotas, which were also used to evacuate wounded troops and prisoners of war. Operation 'Varsity' was the last Allied airborne operation on this scale of the Second World War.

Dakota N473DC, operating out of the Lincolnshire Aviation Heritage Centre, has an impressive Second World War history, which includes preparations for Operation 'Varsity'. The aircraft regularly takes part in air shows and commemorative events and provides taxi rides.

The Middle East

When Japan declared war on Britain and the United States through its attack on Pearl Harbor on 7 December 1941, it became immediately evident that China was to play a key role in the conflict. It was important that the Allies kept China secure in order to use its airbases as a stepping-stone towards a possible attack on Japan. It soon became apparent that Japan was intent on claiming China for itself, initially by compromising China's supply lines, both from the sea and by rail. The only way to keep China supplied was by air and, in order to do this in reasonable safety, the route taken would be via Burma and over part of the Himalayas. The task was given first to the US 10th Air Force and then to the USAAF's Air Transport Command.

The plan was to offload cargo at Indian ports and then transport it by land to Assam, where

BELOW Flt Lt Lord in the fuselage of a Dakota during a supply drop over Burma. Note the two machine guns, mounted where the escape hatches should be. *(RAF Museum)*

LEFT N473DC and N1944A flying together. *(Richard Paver)*

Dakota survivor – N473DC

December 1943 – Delivered to the USAAD with USAAF serial number 42-10082. Assigned to the 9th Air Force Troop Carrier Command (ETO).

February 1944 – Flew to the UK and assigned to the 87th Troop Carrier Squadron, 438th Troop Carrier Group IX Troop Carrier Command, RAF Greenham Common (438th Group provided the four lead squadrons on D-Day, seen off to France by General Eisenhower himself). Fitted with model 80 glider pick-up gear to recover an assault glider from the D-Day landing grounds for refurbishment and reuse in future operations.

July/August 1944 – Towed an assault glider in Operation 'Elmira' on D-Day (carrying soldiers of the 82nd Airborne Division) and resupplied after the landings. Recovered WACO assault gliders from Normandy landing grounds.

2 September 1944 – Transferred to the RAF and given serial number TS422.

September 1944 – Assigned to No.1 Heavy Glider Servicing Unit, 38th Airborne Forces Group, RAF Netheravon. Recovered gliders from Normandy landing grounds and collected and delivered WACO Hadrian and Airspeed Horsa assault gliders to RAF bases prior to Operation 'Market Garden' (Arnhem) and later, in March 1945, in readiness for Operation 'Varsity', the Rhine Crossing.

August 1945 – Transferred to 435 (RCAF) Squadron at RAF Down Ampney under RAF control. Carried out freight duties and troop movements throughout Europe and Scandinavia, including returning Allied POWs.
During the Second World War the aircraft suffered battle damage from various operations. Damage included numerous bullet holes from aerial attack to cockpit and nose areas, including a bullet through the back of the pilot's seat, flak and ground fire damage to under-wing and rear fuselage areas. Dates and locations of damage are unknown. Crew injuries in action are also unknown;

however, casualties would most likely have related to the Arnhem resupply campaign.

April 1946 – Given to the RCAF, retaining the serial number TS422. Returned to Canada with other ex-433 Squadron aircraft. Operated throughout Canada in numerous roles, including glider tug, parachute transport, search-and-rescue and medevac. During this period the aircraft was fitted with skis and jet take-off rockets. She also operated in the far north of Canada, well into the Arctic Circle, using a specially fitted sun compass due to the proximity of the North Pole.

1966 – Transferred for civilian use and reclassified as CF-KAZ and later C-FKAZ. The aircraft operated as a passenger airliner and commercial freighter with, among others, Kier Air Transport (1966–70), Trans-Provincial Airlines (1970–78) and Pacific Coastal Airlines (1978–85).

1985 – Sold to the USA and stored in Texas. Reclassified as N58318.

1995 – Used as a freighter for McNeely Air Charter, operating in the southern United States, and based in Memphis, Tennessee.

2002 – Sold to the Arizona Warbirds Museum. Stored in Tucson, Arizona, and restored by Edwards Worldwide Aviation.

2004 – Sold to the current owner, Dakota Heritage Inc. Trustee.

May 2005 – Flown 5,500 miles from Arizona to Liverpool. Route taken: Tuscon–Arizona–Great Falls, Montana–Saskatoon, Saskatchewan, Canada–Churchill, Manitoba–Iqualuit, Baffin Island–Sonderstrom Airbase, Greenland–Reykjavik, Iceland–Wick, Scotland–Liverpool, England. The journey took 34.5 hours.

Winter 2005/6 – Aircraft restored in the D-Day markings of the 87th Troop Carrier Squadron (TCS), USAAF.

2006 – Reclassified N473DC and operational.

it would be flown into bases in southern China and taken by barge across the country. This was successful until 8 May 1942, when the Japanese took Rangoon and cut Allied access to the Burma Road. The aerial route between Assam and Kunming was then put in place. Both British and American pilots flew this route on a regular basis and it became known as 'The Hump', a reference to the Himalayas that had to be crossed during the journey. The height of the mountains caused fluctuating weather conditions; low cloud and visibility, monsoon weather, strong winds (125 and 200mph) and ice were all commonplace occurrences and often they would be in combination.

A variety of aircraft were considered for the operation and finally four were chosen with the weather, altitude and distance in mind: the Dakota C-46, C-87, C-109 and C-54. Initially, the route was flown solely by the Dakota and its variants; however, it was soon found that the Dakota could not manage the heavy cargoes combined with the altitude necessary to climb over the mountains. Instead they flew a circuitous route through the mountain passes, which was dangerous when combined with low visibility and the possibility of ice clouds that they could not fly above. This led to the other aircraft being tried and used, although even they struggled with the harsh weather conditions and there were many losses.

RIGHT Among the more unusual items to be transported in a Dakota was this bullock, 10 June 1944.
(RAF Museum)

BELOW Maintenance being carried out inside the nose.
(RAF Museum)

BELOW RIGHT Spray-painting the flying controls in the field during the Second World War.
(RAF Museum)

Working The Hump was a difficult task, especially when many of the pilots were inexperienced and there was a shortage of navigational equipment. Navigators and some of the crew would do three round trips in a day. There was also the risk of being attacked by Japanese fighters, a not uncommon occurrence. At one point at least 50 per cent of aircraft in use were being lost each month.

The engineers suffered, too. They worked incessantly outdoors to service and repair the aircraft, which left them vulnerable to sunburn and sunstroke. They were also required to trek into the mountains to cannibalise crashed aircraft for valuable spares. Conditions improved slightly as time went on, yet The Hump is still regarded as one of the hardest trips to make in an aircraft.

The operation began in April 1942 and from May it continued on a daily basis until August 1945, with the last run taking place in November 1945. It was a lifeline to the Chinese, just as the Berlin Airlift was to the Berliners.

LEFT Engine
maintenance being
carried out; note
that the aircraft is in
polished aluminium.
(RAF Museum)

VIP transportation

The Dakota proved to be one of the most popular transportation choices of the Second World War; this didn't just apply to the movement of troops and supplies, but also to both military and civilian VIPs.

King George VI used General Maitland Wilson's personal Dakota, Mk III FZ631, when visiting troops on the Italian front in July 1944. The aircraft was suitably named 'Freedom'. The King requested a Dakota – Mk IV, KN386 – when he travelled with Queen Elizabeth to the Channel Islands in June 1945.

A personal Dakota was frequently modified to the individual's specifications, which would vary according to use and preferences. One of the most luxuriously fitted aircraft was that of Lord Tedder, the Chief of Air Staff. His Dakota Mk IV, KJ994 'Dulcie', was carpeted, panelled in walnut, and featured cream pigskin overhead on the cabin ceiling. The

seating comprised five leather armchairs and two settees, which also doubled as sleeping accommodation, and each interior light could be individually operated. A folding table allowed food, freshly cooked in a small galley, to be served, to the accompaniment of music from a radio set in the main cabin. The height of luxury, indeed.

Field Marshal Montgomery also retained a Dakota for personal use and, after losing one to enemy action, he was given another by General Eisenhower, which was sent to the RAF Maintenance Unit at Courtrai to be modified according to Montgomery's requirements. Although Dakota Mk IV, KN628, was heavier than normal as a result of its embellishments, the performance was adequate. The interior was similar to Lord Tedder's, but without the provision of a galley or sleeping accommodation. The crews of these personal aircraft were usually seconded for a period of time, or occasionally specific personnel were requested.

The Berlin Airlift

After its defeat in the Second World War, Germany was divided between Great Britain, France, the United States and the Soviet Union. The division also applied to Berlin where half the city was under Soviet rule, with the remainder being held by the other nations.

Despite the forming of the Allied Control Council, a provisional government consisting of members from each country with the specific aim of controlling and rebuilding Berlin, the Soviet Union wanted total control of Berlin. It disagreed with many of the council's changes and in order to gain sole control began to blockade the city, disrupting supply lines, in an attempt to force the other countries out. On 9 April 1948 Stalin ordered all American military personnel to leave the Eastern Zone. He began stopping trains on 1 June and by 24 June all land and water access was barred.

There were 2,008,943 Berliners who required the means to eat, keep warm and conduct everyday life. There were also the additional occupation forces to consider and, after tentative diplomatic advances failed, the Allied commanders met to plan a way to overcome the situation. US Military commander General Lucius D. Clay suggested sending an armed convoy through Soviet-controlled Germany to punch its way through the blockade, however, due to the sensitive nature of the situation this could have brought about a Third World War and was not pursued.

British commander, General Sir Brian Robertson, had his own idea: a non-aggressive plan to supply the city by air. Three years before, on 30 November 1945, the Soviet Union had agreed that there would be three 20-mile air corridors, which would provide access to the city. The Soviets had no reason to attempt to block the airways if they were being used for non-military means, and so it was decided that with the Dakota and the C-54 the Berlin Airlift would begin. Codenamed Operation 'Vittles' by the Americans and Operation 'Plane Fare' by the British, the first aircraft began to fly into the city while the Soviets sat back and waited for Berlin to fall.

Brigadier General Joseph Smith was appointed Task Force Commander of the airlift and on 26 June the first Dakota landed at Tempelhof Airport. After due consideration, the city's daily food ration was determined to be:

BELOW RAF Dakotas preparing for the Berlin Airlift.
(RAF Museum)

19 tons of powdered milk, 5 tons of whole milk (for the children), 180 tons of sugar, 11 tons of coffee, 3 tons of fresh yeast, 144 tons of dehydrated vegetables, 38 tons of salt, 10 tons of cheese, 646 tons of flour and wheat, 125 tons of cereal, 64 tons of fat, 109 tons of meat and fish and 180 tons of dehydrated potatoes. A total of 1,534 tons were needed to provision Berlin's population every day.

As well as food supplies, the city required vast amounts of coal and fuel. Heating and the means to cook were important, but industry also required coal. With the Dakota able to carry a limited supply of only 3½ tons, and no sign that the Soviets were relenting, the Allies realised that only experienced, expert organisational skills would give the operation any chance of success.

Although Brigadier Smith had organised the flights on a 'block system' of three eight-hour shifts of a C-54 section followed by a DC-3 section, with aircraft taking off every four minutes, it was obvious that in the long term this would not be enough. On 28 July 1948 Maj-Gen William H. Tunner took over the operation. Drawing on his previous experience gained during the airlift over The Hump into

China in 1944–45, Tunner quickly identified and rectified the weaknesses in the Berlin operation, increasing its effectiveness and gaining the nickname 'Tonnage Tunner'. His first task was to transfer a further 72 C-54s to Wiesbaden and Rhein-Main Air Base, to add to the 54 already flying on the airlift.

On Friday 13 August Tunner flew to Berlin to give Lt Paul O. Lykins an award for the most

ABOVE All sorts of freight was carried inside the DC-3, including rolls of newsprint. *(RAF Museum)*

LEFT Cases of cigarettes are loaded into a Dakota at RAF Lubeck to take to Berlin. *(RAF Museum)*

flights into Berlin to date. With heavy rain and low cloud cover, Tunner's aircraft was held in a stack, circling with many other aircraft. Three aircraft had accidents on landing: one burnt, one burst its tyres trying to avoid the first and the third aircraft ground-looped, blocking the runway and closing the airport. Tunner ordered all aircraft to return to base, realising that he had to devise a more effective system.

The new method depended on the pilot flying according to IFR (Instrument Flight Rules), referring only to instruments, and if the aircraft failed to land on its first attempt it would return to base without delay. There was an immediate change in the efficiency of the operation.

Tunner also identified another weak link: the Dakota. With its sloping cargo floor, which made loading onto a truck difficult, the Dakota took just as long to unload as a C-54, yet it only carried 3½ tons compared to the C-54's 10 tons. The Dakota was taken off the airlift operation, having made a considerable and valiant contribution. Tunner went on to further modify the system, increasing landings to 1,440 per day, which equates to an aircraft landing every three minutes around the clock.

The blockade ended on 12 May 1949; however, the airlift continued, in order to build up supplies and therefore the airlift officially ended on 30 September 1949. During these

15 months the British had delivered 541,937 tons and the Americans 1,783,573.7, more than 2.3 million tonnes in total.

The Airlift Memorial at Rhein-Main Air Base, Frankfurt, and the Luftbruckenplatz at Tempelhof Airport, Berlin, commemorates the pilots and crews that flew during this operation and remembers the 101 lives that were lost.

Berlin Airlift's 30th anniversary – David Peet

David Peet was the pilot of ZA947 from 1978 until 1980. At that time the Dakota was owned by the Royal Aircraft Establishment and based at Farnborough. One of his fondest memories is of flying what is now the BBMF's Dakota out to Tempelhof for the 30th anniversary celebrations of the Berlin Airlift.

One of my best memories of the Dak is taking her to Berlin, on static display, for the celebration of the 30th anniversary of the Airlift, in May 1979. Just flying into Tempelhof evoked many images. I resisted the temptation to throw sweets out the DV window on finals!

We were there for three days, hosted magnificently by the USAF and the citizens of

Berlin. We parked the aircraft under the canopy of the fantastic pre-war terminal. That building was so impressive; the steel pillars holding the massive curved cantilever roof were only about 4ft in diameter and still in perfect condition. Apparently the Nazi plan was to seat some 70,000 people on the roof to watch propaganda spectaculars, but it was never completed. We were also told that there were six storeys below ground with lifts large enough to carry a fighter aircraft to the factory and maintenance facilities below. We wanted to see for ourselves, but were told that it was now too dangerous, so I don't know if that was really true.

The hugely impressive venue was completely overshadowed by the people who visited over the two-day event. We had a constant stream of West Berliners visiting the aircraft who were fanatically interested in the airlift and the aircraft. Their knowledge of both put us to shame. The one question they all asked was 'Did this aircraft participate in the airlift?' I wish I could have told them that it had, but I had to confess that she was still with the RCAF in Canada then. This event was staged every other year and funded by the citizens of West Berlin. Aircraft flew in from many places, some even from the States. We were treated like royalty, almost as if we had flown the airlift ourselves (I was five at that time).

One of the most poignant conversations was with an English lady, probably in her 50s, with her grown-up son. She said, 'I see that aircraft has Royal Aircraft Establishment painted on the fuselage, is that still Farnborough?' I confirmed that it was, and asked her if she had some connection. It transpired that she had worked in the parachute section at Farnborough during and after the war. Her fiancé was a flight engineer on Dakotas and was one of the (surprisingly) few fatal casualties on the airlift. She was pregnant with their son at the time, but because they weren't married she had no entitlement to a RAF widow's pension. After the airlift ended, the people of West Berlin set up a fund for military personnel who had suffered in their cause. She had been paid a handsome pension for life and her son had been put through private school and university. They were both flown out to Berlin for each biennial occasion and treated as VIPs. How's that for gratitude?

The day we departed, we were officially asked to exercise our rights, under the Four Nation Partition Treaty, to fly over the east of the city. We took great delight in exercising our rights. This was, of course, before détente and perestroika and they were still very much 'the enemy', so this was a great PR opportunity not to be missed. It was all filmed from an AAC chopper and the photos were all over the West Berlin papers the next day.

On our way home, we flew through some pretty awful weather and had to don our RAF raincoats – back to front! We always kept a couple on the plane, even in summer, as the ground crew could never seal the DV windows effectively and the rain came in with great force.

Another great trip, which came up twice a year, was to fetch explosive powder (I kid you not – it was in large cardboard drums) from Bremgarten in southern Germany, to Warton. It then went under escort by road to Chorley for making MRCA trials ammunition. We usually managed to arrange departure from Farnborough on a Friday. Warton closed at the weekend, so we had to kick our heels for a couple of days on allowances in the lovely old city of Freiburg.

One Monday morning in the middle of winter when we came to start, we found the battery was flat. The Germans didn't have a starter trolley. We did have an APU (auxiliary power unit) fitted in the toilet compartment, but I didn't dare use it on this occasion because I knew it ran hot and spat sparks, and we had a couple of tons of explosive aboard. With great trepidation, I got the hand starter out of the stowage. I'd never used it and had little faith in my ability to start the engine with it. The aircraft had been standing outside for three days in near zero temperatures. Now, if you've ever started a Dak, as a single pilot, you will know that you need at least three hands. Fortunately, Colin Forrester, my navigator, had always shown a great interest in what his pilot was doing and enjoyed flying the plane. I briefed him to 'mesh' the starter when I nodded my head, then 'boost and prime'. Fully anticipating another couple of nights in Freiburg awaiting spares, I fastened the hand starter into the starboard engine and started to turn the handle. It was really hard and after about 15 seconds I had got as much

ABOVE RAF Dakotas of 1325 Flight, KN958 and KN434, en route from Christmas Island to Changi in 1959. These aircraft are in the RAF Transport Command colours of grey and white with a blue lightning stripe. *(RAF Museum)*

energy into the flywheel as I could muster, and nodded my head to Colin. No one was more surprised than I was when the motor spluttered and coughed into life. I got a round of applause from the onlookers.

During my time at Farnborough, I flew the Red Devils' Islander for their free-fall displays. When their boss, Captain Mickey Munn, heard that I was flying the Dak he asked if we could borrow her to drop the team at Arnhem for the annual celebration. I thought this was a fantastic idea and tried to get approval. The stumbling block was someone in the MOD [Ministry of Defence] who required us to get Handling Squadron at Boscombe to clear the aircraft for dropping parachutists. How many hundreds of thousands of paras have jumped out of Dakotas? Sadly, we weren't able to organise this in time, so a great opportunity was missed. As I had dropped thousands of paras, it would have been a wonderful trip at the end of my

RAF career. I believe it took some years to gain approval for dropping.

The avid enthusiast

The Dakota has many enthusiasts across the world. They include those who have depended on the aircraft in a combat scenario, such as the Second World War and Vietnam, and those who admire the aircraft for her versatility and individual style. Over the years the DC-3 has gained a number of nicknames, many associated with its purpose at that time. Examples include: Dakota (in England), Dumbo, Old Fatso, Old Methuselah, The Dragon Ship and the Gooney Bird. So popular was the name Gooney Bird (believed to be after the large albatross of the South Pacific and its power of flight, specifically its ungainly, but safe landings) that Oscar Brand wrote a song with the same title about the aircraft.

The Gooney Bird song

In '51 they tried to ground the noble DC-3
And so some lawyers brought the case before the CAB
The board examined all the facts behind their great oak portal
And then pronounced these simple words, 'The Gooney Bird's immortal'

Chorus
They patch her up with masking tape
With paper clips and strings
And still she flies, she never dies
Methuselah with wings

The Army toasts their Sky Train now in lousy scotch and soda
The Tommies raise their tankards high to cheer their old Dakota
Some claim the C-47's best, or the gallant R4D
Forget that claim, they're all the same, the noble DC-3

Chorus

Douglas built the ship to last, but nobody expected
The crazy heap would fly and fly no matter how they wrecked it
While nations fall and men retire and jets get obsolete
The Gooney Bird flies on and on, at 11,000 feet

Chorus

No matter what they do to her, the Gooney Bird still flies
One crippled plane was fitted out with one wing half the size
She hunched her shoulders, then took off, I know this makes us laugh
One wing askew, and yet she flew . . . the DC-2 and a half

Chorus

She had her faults, but after all, who's perfect in this sphere?
Her heating system was a gem, and we loved her for her gear
Of course, her windows leaked a bit when the rain came pouring down
She'd keep you warm, but in a storm it's possible you'd drown

Chorus

Well now she flies the feeder routes and carries mail and freight
She's just an airborne office or a flying twelve-ton crate
They patch her up with masking tape
With paper clips and strings
And still she flies, she never dies
Methuselah with wings

'The Last Time' – Trevor Morson

Across the world regular events are held to celebrate these indefatigable aircraft and 2010 brought a special anniversary for the Dakota. Trevor Morson, webmaster and website owner of the DC-3 Hangar (www.douglasdc3.com), was fortunate to attend the 'The Last Time', which saw 26 Dakotas gathered together to celebrate the 75th anniversary of the aircraft. Trevor gives us a taste of the three-day event.

'The Last Time' was an event that no DC-3 enthusiast would want to miss. I am lucky to live in Aurora, Illinois, and Whiteside Airport in Rock Falls, where the event was held, is only an hour's drive west for me. It was going to be the largest gathering of DC-3s in history, celebrating the 75th anniversary of this venerable aircraft.

With media credentials and VIP tickets in hand, representing my website, the DC-3 Hangar, my girlfriend Patti and I were very excited to be a part of this event over the next three days.

Just a mile or so out, before arrival, we were blessed to see a C-47 in Second World War D-Day markings fly right across us at only a few hundred feet above 1-88 highway. With the sound of those humming Pratt & Whitney engines, this was truly our own personal welcome and the rush of adrenalin it provided marked our arrival much like a salute in a perfect way. It was a sight to behold and it will be an everlasting memory for me.

We first checked into our hotel, a few minutes away from the airport, and it was evident that the entire community was supportive and excited to be part of this historic celebration. With everybody wearing DC-3 T-shirts and signs, flyers and billboards with 'Welcome DC-3 enthusiasts' everywhere, along with a couple of DC-3s now flying over us in formation, we quickly realised that we had just arrived in DC-3 heaven!

Our arrival at Whiteside Airport on the morning of Saturday 24 July 2010 was no disappointment. We could see many DC-3s parked everywhere and it was 'all access'. Our first chore of the day was to find Dan Gryder, chief pilot of Herpa DC-3 N143D. I had worked with Dan in email about this event, and have known him, over the Internet, for many years. This was to be our first ever meeting face to face. Surprisingly, Dan was easy to find and, although he was very busy as the main organiser of the event, we found the time to meet and chat for a while.

Throughout the next three days we embraced the sights and sounds of 26 DC-3s and a DC-2. We spoke with many DC-3 pilots – we would gather in the VIP hangar, sharing stories and discussing de-rated power take-off (yes, this controversial DC-3 discussion still exists between us all; 44 inches of manifold

BELOW Twenty-six DC-3s meet at Rock Falls, Illinois, for a week-long celebration of the aircraft. *(Trevor Morson)*

pressure, 46 inches or full power at 48?). This was a fun topic and we often laughed about it. Steve Dunn, pilot of Eastern Airlines DC-3 N18121, used 52 inches of MAP.

I have always found it interesting how music and aviation seem to go hand-in-hand; the evening found us being entertained, sitting under a DC-3 in the VIP hangar, by pilots playing songs by request.

The second day and more DC-3s arrived and it was a perfect day for me to have my friend Jim take me up in his Mooney aircraft for some of the best aerial photography shots one could ever

take. It would be the only way you get each of the 26 DC-3s (and a DC-2) in one photograph.

On the third and final day all of the DC-3s were to take off and fly in formation, heading north to EAA's Oshkosh Air Show. This was a fantastic sight to see and made for another everlasting memory for all those thousands of people who attended the event; volunteers, supporters and enthusiasts alike, we had all just witnessed history before our very eyes.

'The Last Time' event was indeed DC-3 heaven, but I am hopeful that this is 'The First Time' of many more DC-3 gatherings to come.

Into the future

The Dakota has been such a reliable workhorse that today there is a company in the US, Basler Turbo, that converts the DC-3 into a more efficient and reliable aircraft by replacing the original engines with Pratt & Whitney PT6A-67R turbine engines and five-bladed metal propellers.

The fuselage is stretched 40in forward of the wing and the cockpit bulkhead is moved forward a further 60in to increase the transportation capacity of the aircraft. The outer wing leading edge and the wing tip have been redesigned to give better performance and long-range fuel tanks have been incorporated into the outer wings, plus all the flying controls are now metal rather than fabric covered.

The cockpit instrumentation has been modernised and resembles the cockpit of a modern airliner. Also updated is the electrical system, with a complete aircraft rewire, new generators and even an electrically heated, one-piece windscreen. The fuel system has been brought up to date with new electrically driven fuel and transfer pumps and a computerised fuel and flow indication system.

These modified aircraft are employed by a number of air forces around the world, such as the Columbian Air Force and the Royal Thai Air Force. They are also used for environmental research, firefighting and Arctic and Antarctic operations.

When Donald Douglas sat down in front of his drawing board with pencil in hand, I wonder whether he realised that his design would comfortably fly into the 21st century.

Chapter Two

Anatomy of the Dakota

From an engineer's point of view, the Dakota is a well-constructed, easy-to-work-on aircraft. Most of the systems are straightforward to access, the internals of the wings are simple to inspect due to the amount of quick-release fasteners on the panels, and the technical manuals for the aircraft are well written. In this chapter we will get 'under the skin' of the aircraft itself and its many systems.

OPPOSITE Centre section to outer wing rib, showing the boundary angle that bolts the centre section to the outer wing. *(Paul Blackah/Crown Copyright)*

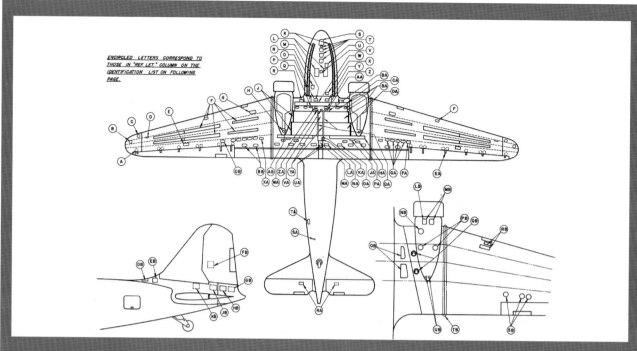

ENCIRCLED LETTERS CORRESPOND TO THOSE IN "REF LET." COLUMN ON THE IDENTIFICATION LIST ON FOLLOWING PAGE.

ABOVE Inspection panel chart. *(AN-01-40NC-2)*

Dimensions of the DC-3 Dakota

Wing span
  95ft
Length
  (with tail cone)
  64ft 5½in
  (with glider tow)
  63ft 9in
Height (tail wheel
  on ground)
  16ft 11.125in
Tailplane span
  26ft 8in
Fuselage width
  8ft 1.125in
Fuselage height
  8ft 8½in

RIGHT Dimensions and areas diagram. *(AN-01-40NC-2)*

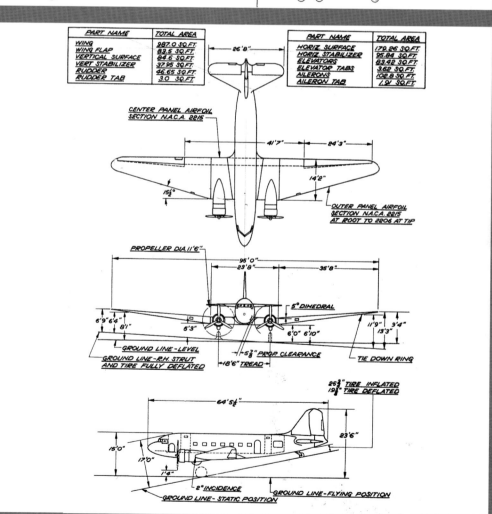

PART NAME	TOTAL AREA
WING	987.0 SQ.FT.
WING FLAP	83.5 SQ.FT.
VERTICAL SURFACE	84.6 SQ.FT.
VERT STABILIZER	37.95 SQ.FT.
RUDDER	46.65 SQ.FT.
RUDDER TAB	3.0 SQ.FT.

PART NAME	TOTAL AREA
HORIZ. SURFACE	179.26 SQ.FT.
HORIZ. STABILIZER	95.84 SQ.FT.
ELEVATORS	83.42 SQ.FT.
ELEVATOR TABS	3.62 SQ.FT.
AILERONS	102.8 SQ.FT.
AILERON TAB	1.91 SQ.FT.

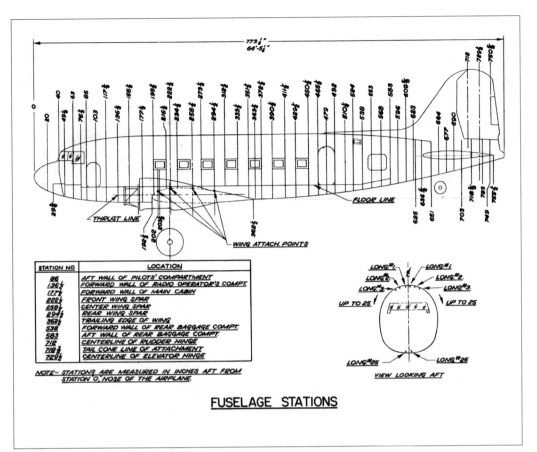

LONG#1. LONG#1.
LONG#2. LONG#2.
LONG#3. LONG#3.

UP TO 25   UP TO 25

LONG#25   LONG#25

VIEW LOOKING AFT

STATION NO.	LOCATION
86	AFT WALL OF PILOTS' COMPARTMENT
136½	FORWARD WALL OF RADIO OPERATOR'S COMPT.
177½	FORWARD WALL OF MAIN CABIN
222½	FRONT WING SPAR
258½	CENTER WING SPAR
294½	REAR WING SPAR
362½	TRAILING EDGE OF WING
538	FORWARD WALL OF REAR BAGGAGE COMPT.
583	AFT WALL OF REAR BAGGAGE COMPT.
712	CENTERLINE OF RUDDER HINGE
718¾	TAIL CONE LINE OF ATTACHMENT
722½	CENTERLINE OF ELEVATOR HINGE

NOTE – STATIONS ARE MEASURED IN INCHES AFT FROM STATION 'O,' NOSE OF THE AIRPLANE.

FUSELAGE STATIONS

THRUST LINE

FLOOR LINE

WING ATTACH POINTS

LEFT The fuselage frame stations. *(AN-01-40NC-2)*

BELOW LEFT The open tail hoist panel, showing the cable. This cable is used to lift the rear of the fuselage. *(Paul Blackah/ Crown Copyright)*

BELOW Aircraft manufacturer's data plate showing the date of construction: 14/3/42. It also gives the aircraft type, C47A-60-DL, and serial number AC42-24338. *(Paul Blackah/Crown Copyright)*

Airframe

The DC-3 Dakota is a low-winged, two-engined monoplane aircraft, of all-metal semi-monocoque construction. For the purpose of this book we will examine each major section of the aircraft and its systems individually.

Fuselage

The fuselage is constructed with upright frames and lengthwise stringers complete with some special longitudinal channels. The fuselage is then covered with Alclad sheet riveted into place.

Pilot's compartment

The pilot's compartment is located forward of station 86 (stations are measured in inches aft from the nose of the aircraft, which is station 0). This compartment may be entered through a door in the bulkhead aft of the radio

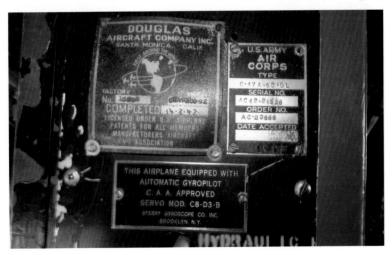

Douglas C-47

Dakota IV cutaway.

(Mike Badrocke)

1 Hinged nose cone, access to instruments and controls
2 Rudder pedals
3 Instrument panel
4 Windscreen de-icing fluid spray nozzle
5 Starboard propeller
6 Windscreen panels
7 Co-pilot's seat
8 Engine throttles
9 Control column
10 Cockpit floor level
11 Access panels to control cable runs
12 Pitot static tubes
13 Aerial cables
14 Propeller de-icing fluid tank

15 Pilot's seat
16 Cockpit bulkhead
17 Cockpit roof escape hatch
18 Whip aerial
19 Starboard landing/ taxiing lamp
20 Windscreen de-icing fluid tank
21 Starboard baggage compartment
22 Electrical fuse panel
23 Crew entry door
24 ADF loop aerial housing
25 Life raft stowage
26 Port baggage compartment
27 Main cabin bulkhead
28 Radio operator's seat
29 Air scoop
30 Heating and ventilating system heat exchangers
31 Astrodome observation hatch
32 Starboard outer wing panel

33 Pneumatic leading-edge de-icing boot
34 Starboard navigation light
35 Starboard aileron
36 Aileron cable controls
37 Trim tab
38 Trim tab control gear
39 Flap control shaft
40 Starboard outer flap
41 Fuselage frame and stringer construction
42 Centre fuselage main frames
43 Centre wing section corrugated inner skin
44 Port main fuel tank, capacity 210 US gal (794 litres)
45 Port auxiliary fuel tank, capacity 201 US gal (760 litres)
46 Wing spar attachments
47 Flap hydraulic jack
48 Centre section flap
49 Floor beam construction

50 Cabin window panels
51 Window panel grommets for small arms attachments
52 Paratroop seating for 28 paratroops
53 Starboard emergency exit window
54 Port emergency window
55 Cabin lining panels
56 Overhead heating and ventilating duct
57 Rear cabin frames
58 Fuselage skin plating
59 Rear cabin bulkhead
60 First aid kit
61 Access door to tail controls
62 Fin root fillet
63 Starboard tail plane
64 Starboard elevator
65 Fin leading edge pneumatic de-icing boot

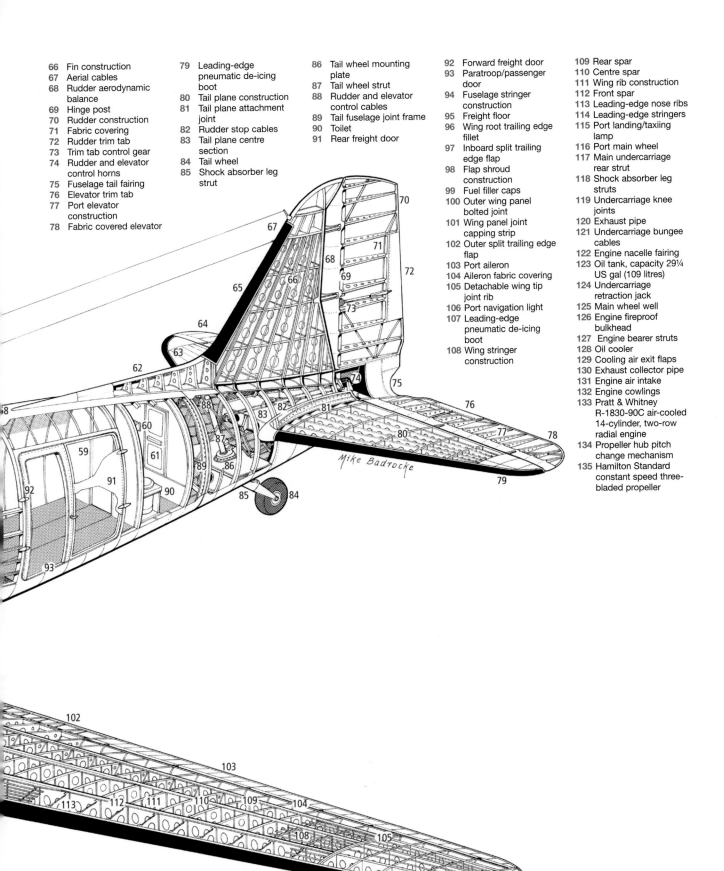

66 Fin construction
67 Aerial cables
68 Rudder aerodynamic balance
69 Hinge post
70 Rudder construction
71 Fabric covering
72 Rudder trim tab
73 Trim tab control gear
74 Rudder and elevator control horns
75 Fuselage tail fairing
76 Elevator trim tab
77 Port elevator construction
78 Fabric covered elevator

79 Leading-edge pneumatic de-icing boot
80 Tail plane construction
81 Tail plane attachment joint
82 Rudder stop cables
83 Tail plane centre section
84 Tail wheel
85 Shock absorber leg strut

86 Tail wheel mounting plate
87 Tail wheel strut
88 Rudder and elevator control cables
89 Tail fuselage joint frame
90 Toilet
91 Rear freight door

92 Forward freight door
93 Paratroop/passenger door
94 Fuselage stringer construction
95 Freight floor
96 Wing root trailing edge fillet
97 Inboard split trailing edge flap
98 Flap shroud construction
99 Fuel filler caps
100 Outer wing panel bolted joint
101 Wing panel joint capping strip
102 Outer split trailing edge flap
103 Port aileron
104 Aileron fabric covering
105 Detachable wing tip joint rib
106 Port navigation light
107 Leading-edge pneumatic de-icing boot
108 Wing stringer construction

109 Rear spar
110 Centre spar
111 Wing rib construction
112 Front spar
113 Leading-edge nose ribs
114 Leading-edge stringers
115 Port landing/taxiing lamp
116 Port main wheel
117 Main undercarriage rear strut
118 Shock absorber leg struts
119 Undercarriage knee joints
120 Exhaust pipe
121 Undercarriage bungee cables
122 Engine nacelle fairing
123 Oil tank, capacity 29¼ US gal (109 litres)
124 Undercarriage retraction jack
125 Main wheel well
126 Engine fireproof bulkhead
127 Engine bearer struts
128 Oil cooler
129 Cooling air exit flaps
130 Exhaust collector pipe
131 Engine air intake
132 Engine cowlings
133 Pratt & Whitney R-1830-90C air-cooled 14-cylinder, two-row radial engine
134 Propeller hub pitch change mechanism
135 Hamilton Standard constant speed three-bladed propeller

Mike Badrocke

ABOVE Cockpit layout of ZA947, the BBMF Dakota. *(Paul Blackah/ Crown Copyright)*

compartment. The cockpit floor is formed from corrugated Alclad sheet, which gives added strength to the floor. An emergency exit is located overhead, between the pilot and co-pilot's seats.

RIGHT The nose bonnet, opened, gives access to the rear of the instrument panel and also the rudder pedals. *(Paul Blackah/ Crown Copyright)*

FAR RIGHT Pilot's seat, removed from the aircraft. The seat is adjustable up and down and forwards and backwards. *(Paul Blackah/Crown Copyright)*

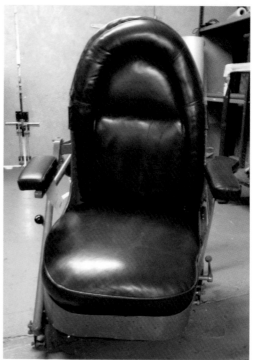

Radio/navigator's compartment

The radio compartment is located between the main cabin and the pilot's compartment, between stations 136½ and 171½. This compartment consists, simply, of a rack of avionics equipment on the starboard side and the navigator's table and seat on the port side. Not all Dakotas are equipped with a navigator's table and this area would then be used as a large luggage space. The BBMF Dakota is fitted with the navigator's table and equipment.

Main cabin

The main cabin extends from the bulkhead aft of the radio/navigator's compartment at station 171½ to the toilet bulkhead. There are seven windows on the starboard side and six windows on the port side of the compartment. The fifth window aft on each side is an escape hatch, which, when operated, swings open allowing personnel to vacate the aircraft in an emergency. There are two types of seating usually fitted to a Dakota: the BBMF Dakota has army-type bucket seats, normally used for para-dropping; other Dakotas may use passenger seating. The para seats can be folded into a stowed position, so it is not necessary to remove them when carrying cargo.

The floor of the compartment is of a

TOP Looking into the cockpit: to the right is the radio equipment rack. *(Paul Blackah/ Crown Copyright)*

ABOVE Rear of the pilot's compartment, to the left the navigator's station. *(Paul Blackah/ Crown Copyright)*

LEFT Main cabin of ZA947 fitted with para seats. These seats fold down to allow the main cabin to carry freight. *(Paul Blackah/Crown Copyright)*

RIGHT **Hinged emergency escape hatch in the closed position.** *(Paul Blackah/ Crown Copyright)*

FAR RIGHT **Main compartment floor is upside down to show its corrugated construction. This construction gives the floor added strength.** *(Paul Blackah/Crown Copyright)*

ABOVE **Main cabin with the floor removed.** *(Paul Blackah/Crown Copyright)*

ABOVE **The main entry door includes a removable para door, which is opened by unlocking the two levers at the top and pulling the door inwards.** *(Paul Blackah/Crown Copyright)*

corrugated construction, again for strength, with cargo tie-down points fitted at various intervals, and is parallel to the line of flight from the forward bulkhead aft to the main cargo-loading door. Here it rises at an angle of 7½° to the aft bulkhead. This rise in the floor provides

a horizontal platform when the aircraft is on the ground and helps in the loading of equipment.

The main loading doors are located on the port side of the fuselage and the forward door has a removable panel, which allows for the dropping of paratroops.

RIGHT **Rear large cargo door in the open position.** *(Paul Blackah/ Crown Copyright)*

FAR RIGHT **Rear fuselage under-floor beams by the cargo door.** *(Paul Blackah/ Crown Copyright)*

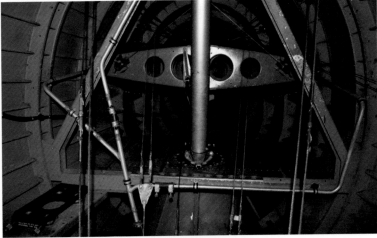

ABOVE Rear fuselage compartment, looking into the stowage area. *(Paul Blackah/Crown Copyright)*

Lavatory compartment

This area is located aft of the main compartment, entered through a door in the rear bulkhead. The toilet is situated on the starboard side and on the port side there is a stowage area for aircraft blanks and chocks.

Tail compartment

The tail compartment is located and entered aft of the lavatory compartment. This gives access to the tailplane centre section and to the tail leg.

Tail cone

The removable tail cone is of an all-metal construction, attached to the aft end of the fuselage.

Tail unit

The tail unit consists of a fin and port and starboard tailplanes. The fin is of aluminium alloy structure, with a removable tip, and is attached to the fuselage with a row of bolts. Two hinge fittings are bolted to the structure for the attachment of the rudder. Two lugs are also provided on the leading edge for the attachment of radio antennae. A de-icing boot can also be installed on the leading edge.

The tailplane is an aluminium alloy structure consisting of two halves (the tailplanes), which are bolted together, complete with a detachable tip. It is then attached to the fuselage by a row of bolts, which are inserted into flange plates riveted to

ABOVE Rear fuselage with tail leg tube assembly, showing control and trim cables, tail-lock cable and de-icing boot air pipes. *(Paul Blackah/ Crown Copyright)*

LEFT Tailcone of the rear fuselage, removed. *(Paul Blackah/ Crown Copyright)*

LEFT Fin, showing the two rudder attachment hinges. *(Paul Blackah/ Crown Copyright)*

RIGHT Fin and rudder diagram. *(AN-01-40NC-2)*

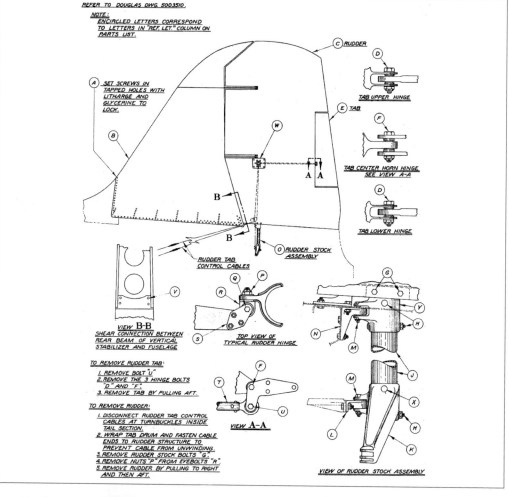

REFER TO DOUGLAS DWG. 5003510.

NOTE:
ENCIRCLED LETTERS CORRESPOND
TO LETTERS IN "REF. LET." COLUMN ON
PARTS LIST.

(A) SET SCREWS IN TAPPED HOLES WITH LITHARGE AND GLYCERINE TO LOCK.

(C) RUDDER
TAB UPPER HINGE
(E) TAB
TAB CENTER HORN HINGE SEE VIEW A-A
TAB LOWER HINGE
(O) RUDDER STOCK ASSEMBLY
RUDDER TAB CONTROL CABLES

VIEW B-B
SHEAR CONNECTION BETWEEN REAR BEAM OF VERTICAL STABILIZER AND FUSELAGE

TOP VIEW OF TYPICAL RUDDER HINGE

TO REMOVE RUDDER TAB:
1. REMOVE BOLT "U"
2. REMOVE THE 3 HINGE BOLTS "D" AND "F".
3. REMOVE TAB BY PULLING AFT.

TO REMOVE RUDDER:
1. DISCONNECT RUDDER TAB CONTROL CABLES AT TURNBUCKLES INSIDE TAIL SECTION.
2. WRAP TAB DRUM AND FASTEN CABLE ENDS TO RUDDER STRUCTURE TO PREVENT CABLE FROM UNWINDING.
3. REMOVE RUDDER STOCK BOLTS "G".
4. REMOVE NUTS "P" FROM EYEBOLTS "R".
5. REMOVE RUDDER BY PULLING TO RIGHT AND THEN AFT.

VIEW A-A

VIEW OF RUDDER STOCK ASSEMBLY

RIGHT Fabric-covered rudder, complete with metal trim tab. The red item at the bottom is the rudder control lock, which is removed before flight. *(Paul Blackah/ Crown Copyright)*

FAR RIGHT Tailplane with fillet panels removed, showing how the tailplane is bolted to the fuselage. *(Paul Blackah/Crown Copyright)*

RIGHT Elevator attachment hinges. *(Paul Blackah/Crown Copyright)*

FAR RIGHT Fabric-covered elevator, complete with two static discharge wicks. *(Paul Blackah/Crown Copyright)*

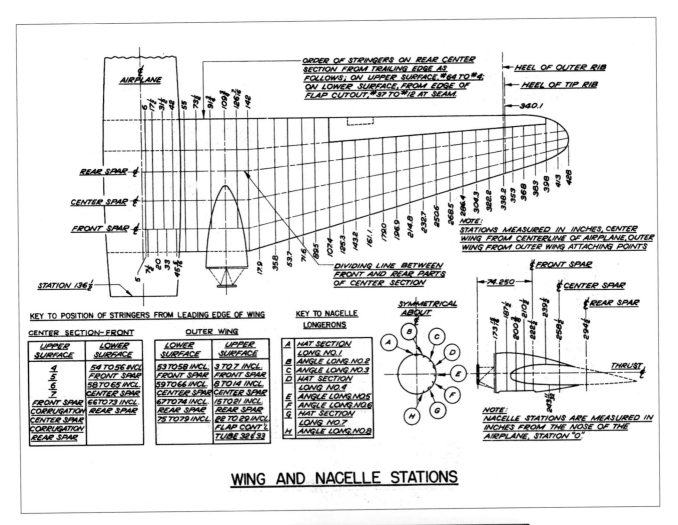

WING AND NACELLE STATIONS

the tailplanes. Four hinge brackets are bolted to the tailplane ribs to enable the attachment of the elevators. On the leading edge of each tailplane a de-icer boot can be fitted. These were fitted as standard, but on some aircraft, such as the BBMF Dakota, they have been removed.

Wing

The wing is an aluminium alloy structure, consisting of a centre section and two outer wing sections. The outer wings are attached to the centre section by numerous steel bolts and lock nuts. A floating rib is fitted between the centre section and each outer wing section to convey the stresses evenly. Each outer wing section has a wing tip attached to it.

Centre section

The centre wing section is constructed with three main spars and an auxiliary spar. The trailing edge section/flap shroud is bolted to the aft spar. Two

ABOVE **Wing frame stations.** *(AN-01-40NC-2)*

LEFT **Outer wing, complete with wing tip and masked-off aileron.** *(Ian Hopwell)*

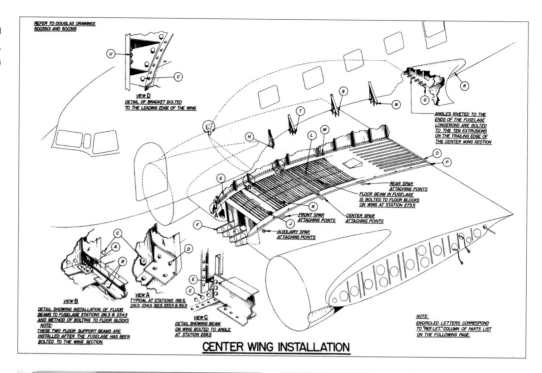

RIGHT Centre wing diagram.
(AN-01-40NC-2)

REFER TO DOUGLAS DRAWINGS 5003501 AND 5003511

VIEW D
DETAIL OF BRACKET BOLTED TO THE LEADING EDGE OF THE WING.

ANGLES RIVETED TO THE ENDS OF THE FUSELAGE LONGERONS ARE BOLTED TO THE TEN EXTRUSIONS ON THE TRAILING EDGE OF THE CENTER WING SECTION.

REAR SPAR ATTACHING POINTS
FLOOR BEAM IN FUSELAGE IS BOLTED TO FLOOR BLOCKS ON WING AT STATION 273.5
FRONT SPAR ATTACHING POINTS
CENTER SPAR ATTACHING POINTS
AUXILIARY SPAR ATTACHING POINTS

VIEW B
DETAIL SHOWING INSTALLATION OF FLOOR BEAMS TO FUSELAGE STATIONS 216.5 & 234.5 AND METHOD OF BOLTING TO FLOOR BLOCKS.
NOTE:
THESE TWO FLOOR SUPPORT BEAMS ARE INSTALLED AFTER THE FUSELAGE HAS BEEN BOLTED TO THE WING SECTION.

VIEW A
TYPICAL AT STATIONS 195.5, 216.5, 234.5, 312.5, 333.5 & 351.5

VIEW C
DETAIL SHOWING BEAM ON WING BOLTED TO ANGLE AT STATION 258.5

NOTE:
ENCIRCLED LETTERS CORRESPOND TO "REF LET." COLUMN OF PARTS LIST ON THE FOLLOWING PAGE.

CENTER WING INSTALLATION

RIGHT Fuselage to centre section, with fillet panels removed. *(Paul Blackah/Crown Copyright)*

FAR RIGHT Leading edge panels removed and opened. To the rear, the fuel tank panel and fuel tank have been removed. The four straps hanging down secure the tank in place. *(Paul Blackah/Crown Copyright)*

RIGHT Port nacelle firewall with the engine removed. *(Jim Douthwaite/Crown Copyright)*

LEFT Inboard end of the outer wing assembly. *(Paul Blackah/ Crown Copyright)*

ABOVE Inside the outer wing, showing the construction of the ribs and frames. *(Paul Blackah/Crown Copyright)*

nacelles are bolted to the centre section and incorporated in each nacelle structure; on the lower 'hat section' are two rubber stops, which the main undercarriage axle rest upon when the undercarriage is retracted. These also provide support to the nacelle structure if a landing is made with the undercarriage retracted. The centre section is bolted to the fuselage by means of eight vertical fittings.

Outer wing section

These are also constructed using three spars. The inboard part of the trailing edge section is attached to the rear spar with nuts and bolts, but on some aircraft this section is riveted in place. On the leading edge of each section a de-icing boot is fitted.

Six brackets are attached to the trailing edge – five on the outer wing and one on the wing tip – for fitting the aileron.

FAR LEFT One of the aileron attachment hinge points, behind which is the aileron input rod, covered by a leather gaitor. *(Paul Blackah/Crown Copyright)*

LEFT Leading edge landing lamp. The three holes in the leading edge are for the pipes to operate the de-icing boots (not fitted). *(Paul Blackah/Crown Copyright)*

LEFT One of the aileron attachment hinge points connected to the aileron. They are locked using two nuts. *(Paul Blackah/ Crown Copyright)*

LEFT Centre section to outer wing joint, shown here with the paint removed. The dark-coloured patch is a steel repair plate. *(Paul Blackah/ Crown Copyright)*

Undercarriage

The hydraulically operated main undercarriage has a fixed, but swivelling, tail leg. Each main undercarriage wheel is mounted on an axle, which is clamped between two air/oil-filled shock absorber struts. The two struts are then connected at their upper ends to rigid trusses. Attached to the bottom of each strut piston tube are two triangular links, the top corners of which are bolted to rods, which are bolted to the shock strut cross bracing. The aft corners of the triangular links are bolted to the yoke-type rear-bracing strut, which is hinged at its upper end to the wing structure.

The upper end of the shock absorbers are hinged to the upper truss, which rotates about its upper fittings, which, in turn, are attached to the front wing spar of the centre section.

A hydraulic jack is bolted to the upper truss and to the nacelle structure aft of the firewall. When the jack is retracted, the upper truss is pulled forward and upwards into the nacelle and, because the shock absorber units, complete with wheel assembly, are hinged to the lower end of the truss, the wheel is also pulled up into the nacelle. When the gear is fully retracted, the projecting end of the axles are held against rubber bumpers, which are built into the sides of the nacelles, so that some of

LEFT Main undercarriage leg complete with the drag stay. (Paul Blackah/Crown Copyright)

BELOW Close-up of the brake unit fitted to the main wheel and axle unit. (Paul Blackah/Crown Copyright)

How to retract the undercarriage

1. Place the mechanical safety latch control handle in the LATCH RAISED position. This will raise the safety latches, releasing the landing gear, actuating cylinder piston rod hooks.

2. Place the landing gear control handle in the UP position. Hydraulic fluid, under pressure, will then be directed to the lower end of the jack, retracting the piston rod, which, in turn, pulls the undercarriage into the nacelles. The green undercarriage light will go out and a red light will come on.

3. When the undercarriage has been fully retracted, place the landing gear valve handle in NEUTRAL position. There are no mechanical locks to hold the gear in the UP position, only hydraulic fluid pressure in the retraction jack holds the undercarriage in the retracted position, therefore the landing gear control valve is placed in the NEUTRAL position after retraction of the landing gear to retain fluid pressure in the upline.

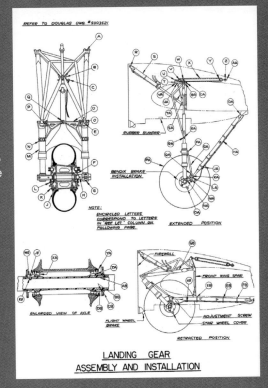

LANDING GEAR
ASSEMBLY AND INSTALLATION

LEFT Landing gear operation diagram. *(AN-01-40NC-2)*

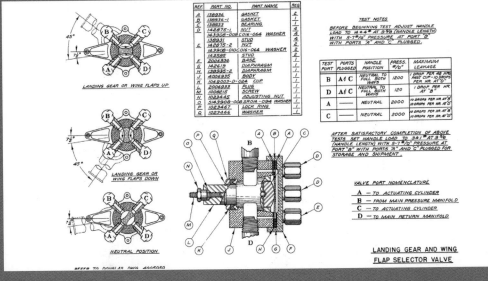

LANDING GEAR AND WING
FLAP SELECTOR VALVE

LEFT Flap and undercarriage selector valve operation diagram. *(AN-01-40NC-2)*

BELOW FAR LEFT To the right is the hydraulic hand pump and to the left is the landing gear latch. Outboard of these levers you can see one of the pilot's and co-pilot's seat rails. *(Paul Blackah/Crown Copyright)*

LEFT Undercarriage selector lever and hydraulic hand pump (lower) on the hydraulic control panel. *(Paul Blackah/Crown Copyright)*

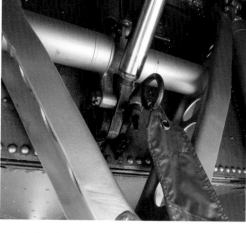

the wheel is seen protruding slightly from the nacelle. A hydraulic compensator jack assists the main hydraulic jack when retracting the undercarriage.

When the aircraft is on the ground, a ground safety pin is fitted to stop the undercarriage from collapsing when hydraulic pressure is dissipated.

Tail leg

The tail leg is fully swivelling and consists of a spindle, fork, wheel and air/oil-filled shock absorbers. The tail-wheel gear is lockable, in the trailing position, by a lever located in the cockpit, below the throttle quadrant, and by cables that run all the way to the tail leg. The tail should be locked for take-off and landing and unlocked for taxiing. The system is designed so that if you forget to unlock the tail leg, a light alloy sheer pin will allow the gear to swing free without causing any damage to the aircraft structure.

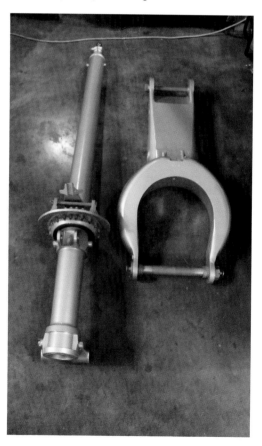

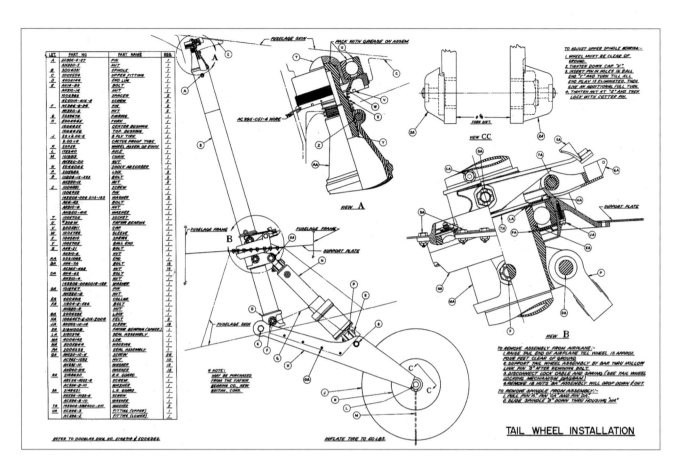

TAIL WHEEL INSTALLATION

Fuel system

ABOVE Tail leg installation diagram. (AN-01-40NC-2)

BELOW Underside rear fuselage, with tail leg removed. (Paul Blackah/Crown Copyright)

The fuel system comprises four fuel tanks constructed from aluminium alloy sheet, which are welded together and strengthened with internal stiffeners and baffles. There are two main fuel tanks and two auxiliary fuel tanks. These are located in the wing centre section. They are supported by a padded cradle assembly and held in place by aluminium straps.

The main tanks have a capacity of 168gal and the auxiliaries 167gal each. Each tank is equipped with an electric gauge connected to an indicator on the instrument panel, a water and dirt collecting sump and a drain cock.

There are two selector valves fitted in the aircraft: the valve for the left system is mounted on the aft side of the centre wing spar and the valve for the right system is mounted on the forward side of the wing spar. These valves are operated by a series of cables and pulleys from their respective controls, which are located on each side of the top control pedestal in the cockpit. These valves select the fuel tank or

RIGHT One of the four fuel tank cradles that support the fuel tank when it is fitted in the centre section. *(Paul Blackah/Crown Copyright)*

FAR RIGHT Inside the fuel tank bay, showing the corrugated construction of the centre section. *(Paul Blackah/Crown Copyright)*

ABOVE One of the welded aluminium-constructed fuel tanks complete with filler cap (red) and contents sender unit (grey). *(Paul Blackah/Crown Copyright)*

ABOVE RIGHT On the starboard side of the main compartment is the internal fuel tank filler (not in use). *(Paul Blackah/Crown Copyright)*

RIGHT Under the floor, just aft of the main compartment front bulkhead, is the internal fuel tanks selector valve (not in use). *(Paul Blackah/Crown Copyright)*

tanks for which fuel is supplied to the engines. In normal operation, the left-hand front or rear tank supplies fuel to the left engine and the right-hand front or rear tank supplies fuel to the right engine. However, any tank may be used to supply either or both engines.

Although not fitted to the BBMF Dakota, long-range fuel tanks can be placed in the main cabin. They are fitted in multiples of two up to a maximum of eight, with each tank having a capacity of approximately 83gal. This system is operated by gravity feed, controlled by a set of four valves beneath the walkway in the main cabin.

Oil system

Each engine has its own independent oil system, with a capacity of 28.3gal. The system comprises an oil tank, located at the top of the nacelle aft of the firewall on the centre section, which holds 24.1gal. The tank is saddle-shaped in appearance to allow clearance for the landing gear retraction jack, which it straddles. Baffles within the tank prevent the oil from sloshing back and forth during flight. A drain cock is provided on the tank sump to allow the contents to be emptied.

An oil cooler assembly is situated aft and

FAR LEFT Aircraft oil tank, with filler point to lower right. *(Dennis Lovesday)*

LEFT Aircraft oil tank. The cut-out in the middle allows for clearance of the retraction jack. *(Dennis Lovesday)*

LEFT Engine oil tank fitted in top of nacelle, with the main undercarriage jack shown in the middle cut-out of the tank assembly. *(Paul Blackah/Crown Copyright)*

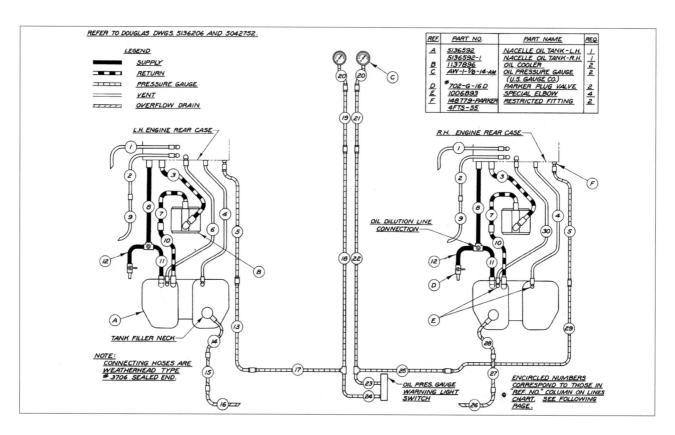

REFER TO DOUGLAS DWGS. 5136206 AND 5042752.

LEGEND
- ▬▬▬ SUPPLY
- ▬◼▬ RETURN
- ⊥⊥⊥⊥ PRESSURE GAUGE
- ═══ VENT
- ▨▨▨ OVERFLOW DRAIN

REF.	PART NO.	PART NAME	REQ
A	5136592	NACELLE OIL TANK–L.H.	1
	5136592-1	NACELLE OIL TANK–R.H.	1
B	1137896	OIL COOLER	2
C	AW-1-⅛-14-AM	OIL PRESSURE GAUGE (U.S. GAUGE CO.)	2
D	#702-G-16 D	PARKER PLUG VALVE	2
E	1006893	SPECIAL ELBOW	4
F	148779-PARKER 4FTS-55	RESTRICTED FITTING	2

L.H. ENGINE REAR CASE

R.H. ENGINE REAR CASE

OIL DILUTION LINE CONNECTION

TANK FILLER NECK

NOTE:
CONNECTING HOSES ARE WEATHERHEAD TYPE # 3706 SEALED END.

OIL PRES. GAUGE WARNING LIGHT SWITCH

ENCIRCLED NUMBERS CORRESPOND TO THOSE IN "REF. NO." COLUMN ON LINES CHART. SEE FOLLOWING PAGE.

ABOVE Engine oil system. *(AN-01-40NC-2)*

below the engine and protrudes into the airflow. The oil cooler is composed of a shell, and outer warming jacket, a core separated by baffles and a mounting flange for the surge valve.

The purpose of the oil system is to lubricate and cool the engine and provide oil, under pressure, for the feathering of the propeller. The system works as follows: oil is drawn from the tank to the engine-driven pump. From here it is forced, under pressure, to lubricate all the parts of the engine. The oil, heated by the engine, flows down into the sump, which is located at the bottom. A scavenge pump picks up oil from this sump and pumps it to the temperature regulator where, under normal conditions, it flows through the cores in the oil cooler, where the heat dissipates into the air stream that flows through the cooler. Operating oil cooler shutters, mounted on the rear of the oil cooler, from the pilot's compartment, can regulate the quantity of air to give the desired oil system temperature.

Hydraulic system

The hydraulic system operates the landing gear, the brakes, wing flaps, cowl flaps and the automatic pilot. Two engine-driven pumps, one mounted on each engine, supply the pressure to operate the system. The pump on the right engine operates the automatic pilot and the pump on the left engine supplies the main hydraulic system; however, there is a selector valve, which allows the pumps to be switched should the need arise. The system operates by fluid passing from the reservoir to both pumps, then on to the engine selector valve. From the engine pump selector, the pump selected to supply the system pressure will provide fluid through the pressure regulator

RIGHT Engine oil cooler. *(Paul Blackah/ Crown Copyright)*

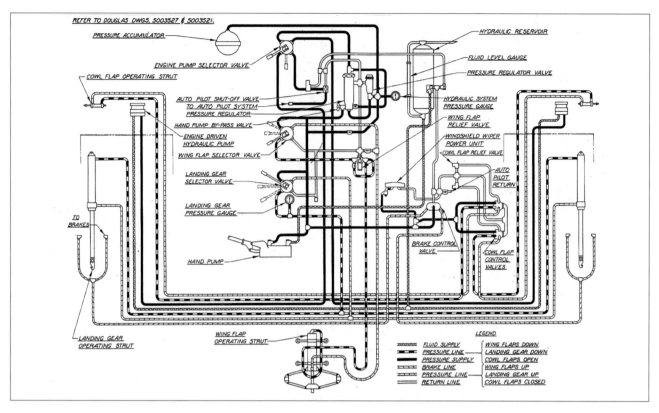

ABOVE **Hydraulic system.** *(AN-01-40NC-2))*

RIGHT **Hydraulic control panel.**
(AN-01-40NC-2)

BELOW **Behind the hydraulic control panel, showing the selector valves and hoses.** *(Paul Blackah/Crown Copyright)*

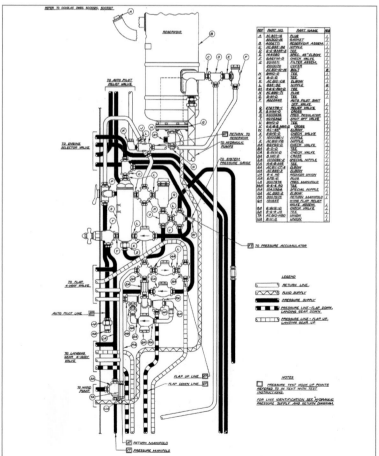

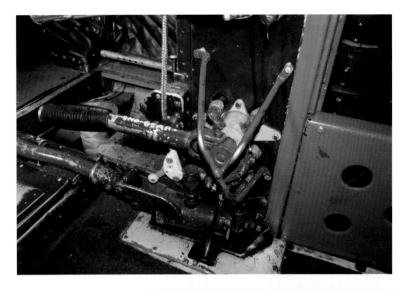

ABOVE Undercarriage selector lever and hydraulic hand pump (lower) on the hydraulic control panel. *(Paul Blackah/ Crown Copyright)*

RIGHT Hydraulic panel showing the hydraulic reservoir, slightly hidden, and the accumulator. *(Paul Blackah/Crown Copyright)*

BELOW Close-up of the overhauled hydraulic reservoir. *(Dennis Lovesday)*

to the hydraulic pressure accumulator, giving an operating pressure of 650–850psi. If the regulator fails and the system pressure exceeds 1,000psi, a system relief valve, adjacent to the regulator, opens and bypasses excess fluid back to the reservoir. From the regulator and accumulator, fluid flows to the hydraulic control panel and from there to the various units in the system. A hydraulic hand pump is installed on the hydraulic control panel, to provide pressure to operate any of the systems in the event of pump failure. The hand pump can also be used for ground operation when the engines are not running in order to check the systems.

The reservoir

The reservoir is mounted behind the hydraulic control panel and is provided with a filler neck, extending out into the walkway, and a sight gauge, which shows the fluid level in the reservoir. The total capacity of the reservoir, when it is filled to the top marker on the sight gauge, is 2½gal; 1¾gal of this fluid is available to the engine-driven pumps and the sump contains ¾gal, which is available only to the hand pump. Approximately 6gal of oil are required to fill the complete system.

Engine-driven pumps

These are Pesco model 214 gear-type pumps and are identical for both engines. The capacity for each pump is approximately 3½gal per minute when working against a pressure of 850psi at an engine speed of 2,100rpm.

Engine pump selector valve

This valve is a tapered plug-type two-way valve with the passages cut in the valve, so that the fluid flow cannot be cut off. The valve connects to the hydraulic pumps, the pressure regulating valve and the automatic pilot system. With the selector handle in the vertical/normal position, the left-hand engine supplies the main system, while fluid from the right-hand pump operates the automatic pilot. Moving the selector handle forward to the alternate position swaps over the operation of the pumps.

Pressure accumulator

This is a forged aluminium spherical housing, divided into two compartments by a rubber diaphragm. Air is contained in the lower compartment at 250psi with no fluid in the upper compartment and no pressure in the hydraulic system. As system pressure builds above 250psi, hydraulic fluid forces the diaphragm down and flows into the upper compartment and when the pressure in that compartment reaches 850psi, it contains approximately 1¼gal of fluid. The purpose of the accumulator is to store energy in the hydraulic system for use when the pumps are inactive. This gives an instantly available supply of pressure fluid in case fluid is drawn from the system faster that the pump can replenish it – for instance, when the landing gear is lowered, the weight of the gear and the air load on the gear causes it to drop rapidly. Without the pressure accumulator the delivery of the pump could not keep pace with the demand of the gear and a vacuum would occur somewhere in the system. The accumulator prevents this from happening by immediately discharging its fluid into the system.

When the gear is down, the engine pump will replenish the accumulator.

Brakes

Two systems allow either brake to be applied independently by the use of brake control units. Applying pressure to the rubber toe pedals transfers the force to the rear end of the lever on the brake control valve, rotating it about its hinges and forcing the brake piston upwards. This action allows fluid, under pressure, to flow to the brake unit therefore applying the brakes. Releasing the toe pedal allows the fluid pressure in the valve, assisted by a spring, to force the mechanism back to the neutral position and so relieves the brake pressure through the line. Each wheel is fitted with a pair of these dual-servo brake units, which are bolted to the main wheel axle unit.

Wing flaps

A valve assembly on the hydraulic control panel operates the flaps. When the lever is placed in the down position the hydraulic fluid is directed from the pressure manifold, through the wing flap release valve to the forward end of the flap actuating jack, which operates the flaps. As the piston in the cylinder moves towards

LEFT Hydraulic accumulator. *(Paul Blackah/ Crown Copyright)*

BELOW Four hydraulic brake units with the two main wheel axle assemblies. *(Paul Blackah/Crown Copyright)*

BOTTOM Flaps shown in the fully-down position. *(Paul Blackah/ Crown Copyright)*

the rear and forces the flaps down, the fluid in the rear of the cylinder flows back through the control valve to the main return manifold. When selected UP, the opposite happens, enabling the flaps to move in an upward direction. A relief valve is fitted into the flap system and if the flaps are lowered or an attempt is made to lower at speeds in excess of 112mph, the flaps will lower until the air pressure against the flaps exerts a pressure equivalent to 400psi at which point the relief valve will operate and the flaps will stop lowering. The flaps will only start to lower again if the aircraft's speed is decreased, enabling the pressure to drop below 400psi.

The cowl gill system

The cowl flaps are hydraulically opened or closed to maintain engine temperature. Each engine has its own system and valves in the cockpit on the starboard side operate them. There are five settings for each valve: OPEN, OFF, TRAIL, OFF and CLOSE. Each engine has a cowl flap-operating strut fitted in the engine

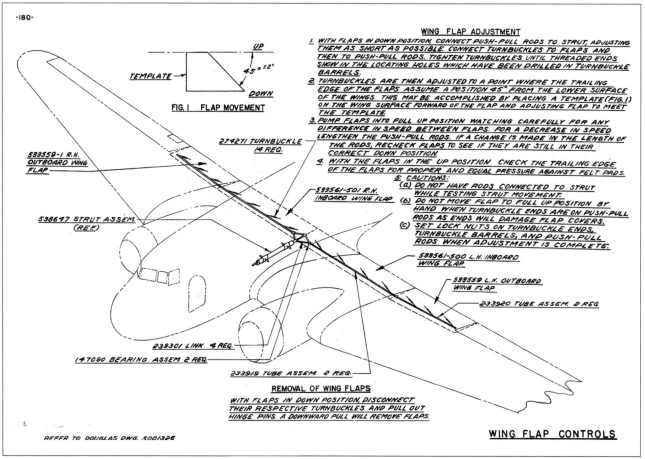

-180-

WING FLAP ADJUSTMENT

1. WITH FLAPS IN DOWN POSITION, CONNECT PUSH-PULL RODS TO STRUT, ADJUSTING THEM AS SHORT AS POSSIBLE. CONNECT TURNBUCKLES TO FLAPS AND THEN TO PUSH-PULL RODS. TIGHTEN TURNBUCKLES UNTIL THREADED ENDS SHOW IN THE LOCATING HOLES WHICH HAVE BEEN DRILLED IN TURNBUCKLE BARRELS.

2. TURNBUCKLES ARE THEN ADJUSTED TO A POINT WHERE THE TRAILING EDGE OF THE FLAPS ASSUME A POSITION 45° FROM THE LOWER SURFACE OF THE WINGS. THIS MAY BE ACCOMPLISHED BY PLACING A TEMPLATE (FIG.1) ON THE WING SURFACE FORWARD OF THE FLAP AND ADJUSTING FLAP TO MEET THE TEMPLATE.

3. PUMP FLAPS INTO FULL UP POSITION WATCHING CAREFULLY FOR ANY DIFFERENCE IN SPEED BETWEEN FLAPS. FOR A DECREASE IN SPEED LENGTHEN THE PUSH-PULL RODS. IF A CHANGE IS MADE IN THE LENGTH OF THE RODS, RECHECK FLAPS TO SEE IF THEY ARE STILL IN THEIR CORRECT DOWN POSITION.

4. WITH THE FLAPS IN THE UP POSITION CHECK THE TRAILING EDGE OF THE FLAPS FOR PROPER AND EQUAL PRESSURE AGAINST FELT PADS.

5. CAUTIONS:
   (a) DO NOT HAVE RODS CONNECTED TO STRUT WHILE TESTING STRUT MOVEMENT.
   (b) DO NOT MOVE FLAP TO FULL UP POSITION BY HAND WHEN TURNBUCKLE ENDS ARE ON PUSH-PULL RODS AS ENDS WILL DAMAGE FLAP COVERS.
   (c) SET LOCK NUTS ON TURNBUCKLE ENDS, TURNBUCKLE BARRELS, AND PUSH-PULL RODS WHEN ADJUSTMENT IS COMPLETE.

UP
45°±2°
DOWN
TEMPLATE

FIG.1 FLAP MOVEMENT

533559-1 R.H. OUTBOARD WING FLAP

274271 TURNBUCKLE 14 REQ.

533561-501 R.H. INBOARD WING FLAP

538647 STRUT ASSEM. (REF.)

533561-500 L.H. INBOARD WING FLAP

533559 L.H. OUTBOARD WING FLAP

233920 TUBE ASSEM. 2 REQ.

233301 LINK 4 REQ.

147090 BEARING ASSEM. 2 REQ.

233919 TUBE ASSEM. 2 REQ.

REMOVAL OF WING FLAPS
WITH FLAPS IN DOWN POSITION, DISCONNECT THEIR RESPECTIVE TURNBUCKLES AND PULL OUT HINGE PINS. A DOWNWARD PULL WILL REMOVE FLAPS.

REFER TO DOUGLAS DWG. 5001325

WING FLAP CONTROLS

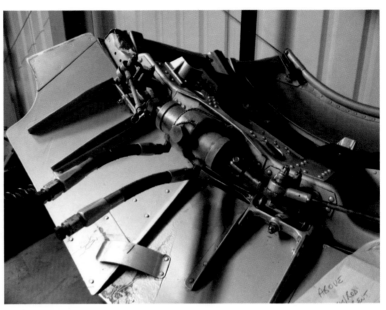

cowlings, which is attached to the cowl flap control tie rods.

Windscreen wiper

A hydraulically driven motor operates the windscreen wipers.

Hydraulic control panel

This unit is situated on the starboard side, just aft of station 86. It contains the hydraulic hand pump, the landing gear control valve, the flap control valve, the hand pump bypass valve, the engine pump selector valve, the automatic pilot

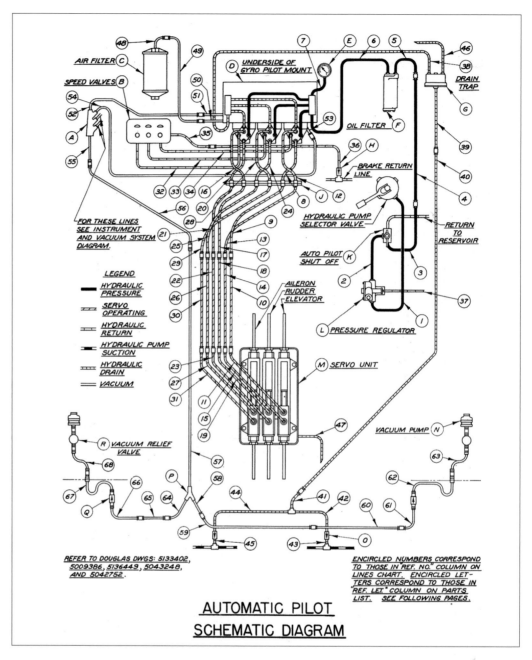

**AUTOMATIC PILOT
SCHEMATIC DIAGRAM**

shut-off valve and the hydraulic reservoir sight gauge and filler neck.

Autopilot

The autopilot system controls the aileron, rudder and elevator systems. The complete installation consists of the following units: directional gyro, bank and climb gyro, mounting unit, oil filter, three servos, speed control units, pressure regulator, air filter, pressure manifold, balanced oil valves and a pressure gauge.

This system has been removed from the BBMF Dakota, because it was found to be unreliable. Many other Dakotas have also had this system removed.

Electrical system

The electrical system can be either 12- or 24-volt, depending on the mark of Dakota. We will take a look at the electrical system of the BBMF Dakota, which is a 24-volt system. The system consists of two generators, operating in parallel, two 12-volt batteries, connected in series, and all the necessary regulating equipment.

Generator specifications	
Minimum speed	2,500rpm
Maximum speed	4,500rpm
Rated voltage	28½ volts

Generators

There are two generators in the system, each one mounted on the rear of the engine and they provide a source of controlled electrical power to the system. The type M-2 generator is designed for direct current and is driven in a clockwise direction. During flight, both generators should be ON unless one generator is defective. The aircraft should never take off with one generator not working because, should a problem arise during flight and the other generator fails, the batteries would then be the only source of electrical power and will only last for a short period of time.

Voltage regulators

The voltage regulators are rubber shock-mounted in the main junction box and they control generator voltage at 28 volts throughout the operating range.

Batteries

There are two 12-volt batteries located in the fuselage bottom, forward of the fuselage wing. Each battery is mounted on a movable platform, which is lowered and raised through inspection panels in the fuselage. The batteries provide current for the electrical systems and are recharged, in flight, by the generators.

The battery master switch

This is located in the main junction box and is remotely controlled by a handle on the floor to the right of the pilot's seat.

Generator current control switch relays

These are used in the generator system to provide circuit control and reverse current protection.

Main junction box

The main junction box is located under the forward end of the navigator's table. It contains

ABOVE Battery stowage with battery in the lowered position. *(Paul Blackah/ Crown Copyright)*

Changing the batteries

Caution: Before removing or replacing the aircraft batteries, always make sure that the battery master switch is in the OFF position.

1. Open the battery access panel doors.
2. Pull out the release handle locking pin.
3. Pull release handle on battery platform straight down until the safety catch on the slide tube engages. *(NB: the safety catch must be engaged before removing the battery, as the platform is spring-loaded.)*
4. Loosen the battery clamps and lift the battery off the platform.
5. Place the replacement battery on the platform and tighten the clamps.
6. Hold the battery platform down and release the safety catch.
7. Push the platform up and insert the locking pin. The required connections to the electrical system are made automatically when the batteries are raised into position, by two spike-like fittings that are attached to the positive and negative terminals on the battery.
8. Close and secure the battery access panel.
9. Turn on the power switch.

the fuses, voltage regulators, reverse current relays, starter relays, landing light relays, dimming relays and the battery master switch.

Lighting

The lighting system consists of pilot's and main cabin lights, fuel and oil pressure warning lights, navigation lights and landing lights. The landing lights are mounted in the leading edge of the wings and provide a beam of light approximately 400ft in front of the nose of the aircraft.

Ignition

The ignition system is of dual magneto type and consists of an ignition switch assembly located between the pilot's left and right

electrical panels, which includes a master switch and a switch for each engine, each of which is equipped with a left and right magneto. A booster system is also included within this system, with the ability to give an additional spark to assist with the engine start. A booster system is attached to each engine.

Left-hand electrical panel

The panel is mounted forward of the pilot's seat, above the left windshield. It contains the compass lamp rheostat, panel lamps rheostat, left propeller feathering switch, battery master switch, landing light switches, two-position light switches, wing and tail passing light switch, propeller de-icer switch, pitot heating switches, oil dilution switches, engine primer switches, the gated bail-out (alarm bell) switch and the paratroop warning light switch.

Right-hand electrical panel

This panel is mounted forward of the co-pilot's seat, above the right windshield. It contains two ammeters, left and right, right propeller feathering switch, two starter switches, energise left and right and mesh left and right, recognition light switches, cockpit light switches and ammeter light switch.

Heating and ventilation

The heating and ventilation system comprises two exchangers, two mixing chambers, four critical temperature warning lights, two critical temperature thermo switches,

two windshield defrosting shields and necessary scoops, ducts, tubing, hose outlet valves and linkages. The system operates by taking air from outside the aircraft, through two muff-type heat exchangers, one mounted on each exhaust tail pipe. The air is heated up by the tail pipes and is routed through the air-mixing chamber, where it meets air that enters through scoops at the belly of the fuselage. The heated air and the non-heated air are mixed, in the necessary proportions, to regulate the temperature of the airflow throughout the aircraft. Air is ducted to the following places within the aircraft: pilots' foot warmer outlets, pilots' hand warmer outlets, windshield defroster outlet, astrodome outlet, navigator's outlet, radio operator's outlet and the main cargo duct outlets within the main cabin. In addition, there are two emergency spill valve outlets, one in each nacelle.

Controls regulating outlet openings are installed in the control box, located on the floor aft of the co-pilot's seat. The five controls, located within the box, regulate the right-hand emergency spill valve, the left-hand emergency spill valve, the left-hand mixing chamber, the right-hand mixing chamber and the cross feed by-pass control. The first three controls have a flexible push-pull linkage to the applicable valves and the other two have rod linkages.

The spill valves each have a critical temperature thermo switch, which light temperature-warning lamps if the temperature exceeds 230°C. These lights are located in the pilot's compartment and the radio-operator's compartment. Immediately, when a warning light indicates, the applicable spill valve should be opened allowing the heated air to escape.

Ice elimination

There are two systems fitted to the Dakota: one system employs mechanical means to break up the ice (known as de-icing), and the other employs either heat or de-icer fluid to prevent ice from forming (anti-icing).

De-icing

The de-icing system works with air from the exhaust of the two engine vacuum pumps flowing through a distributor valve to alternately expand de-icer boots along the leading edges of the wings, tailplane and fin. The inflation and deflation of these boots cracks any ice that builds up along the leading edges. The system is operated by a handle in the pilot's compartment, which turns the distributor valve on and off. This system is no longer fitted to the BBMF Dakota, as she is not allowed to fly in known icing conditions.

Anti-icing

There are two anti-icing systems on a Dakota: the windshield system and the propeller system. The windshield system contains two units, one that sprays alcohol from a supply tank, by

ABOVE LEFT Lagged pipe is for cabin heating. *(Paul Blackah/Crown Copyright)*

ABOVE Aircraft heating control unit, situated behind the co-pilot's seat. Pulling up or pushing down on the levers controls the temperature in both the pilot's and main compartments. *(Paul Blackah/Crown Copyright)*

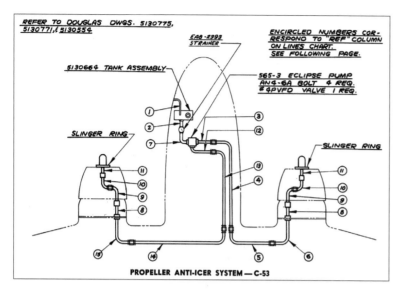

ABOVE **Prop anti-icing system.** *(AN-01-40NC-2)*

REFER TO DOUGLAS DWGS. 5130775, 5130771, & 5130554

EAB-2393 STRAINER

5130664 TANK ASSEMBLY

ENCIRCLED NUMBERS COR-RESPOND TO "REF" COLUMN ON LINES CHART. SEE FOLLOWING PAGE.

565-3 ECLIPSE PUMP
AN4-6A BOLT 4 REQ.
#4PVFO VALVE 1 REQ.

SLINGER RING

SLINGER RING

PROPELLER ANTI-ICER SYSTEM — C-53

3.5gal, through a shut-off valve, filter and electric pump into a slinger ring on the propeller hub. Distribution lines from the ring provide each blade. The pump is turned on or off by a switch on the pilot's electrical control panel. A rheostat, to the right of the tank, regulates the fluid flow by controlling the speed of the electric motor.

Flying controls

The Dakota has dual controls, allowing the pilot and co-pilot to be seated side by side. Each has an individual control column, wheel and a set of rudder pedals to control the main surfaces of the aircraft. In addition, trim tab controls are located on a pedestal between the two seats and are accessible by both crewmembers. The main control surfaces are of metal-frame, fabric-covered construction and the tabs are of all-metal construction. To reduce control loads, all surfaces are aerodynamically balanced. In addition, each elevator has an 8lb counterweight mounted ahead of the hinge line,

BELOW **The control column diagram.** *(AN-01-40NC-2)*

means of a hand pump, to the two right and two left windshields, and the second, which supplies de-icer fluid from the same tank, by means of an electric pump, onto the front windshields. The supply tank holds 5.4gal.

The propeller anti-icing system is designed to prevent ice forming on the propeller blades. Fluid is pumped from the supply tank, holding

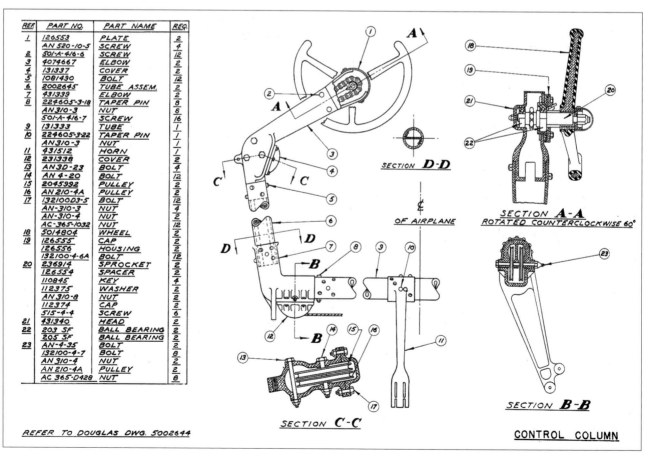

REF.	PART NO.	PART NAME	REQ.
1	126553	PLATE	2
2	AN 520-10-5	SCREW	4
	501-A-4/6-6	SCREW	12
3	4074667	ELBOW	2
4	131337	COVER	2
5	1081490	BOLT	12
6	2002645	TUBE ASSEM.	2
7	431339	ELBOW	2
8	224605-3-18	TAPER PIN	8
	AN310-3	NUT	8
	501-A-4/6-7	SCREW	16
9	131333	TUBE	1
10	224605-3-22	TAPER PIN	1
	AN310-3	NUT	1
11	431512	HORN	1
12	231338	COVER	2
13	AN3D-23	BOLT	4
14	AN 4-20	BOLT	12
15	2045992	PULLEY	2
16	AN 210-4A	PULLEY	2
17	132100D3-5	BOLT	12
	AN-310-3	NUT	4
	AN-310-4	NUT	2
	AC-365-1032	NUT	12
18	5014804	WHEEL	2
19	126555	CAP	2
	126556	HOUSING	2
	132100-4-6A	BOLT	12
20	236914	SPROCKET	2
	126554	SPACER	2
	110845	KEY	4
	112375	WASHER	2
	AN310-8	NUT	2
	112374	CAP	2
	515-4-4	SCREW	6
21	431340	HEAD	2
22	203 SF	BALL BEARING	2
	205 SF	BALL BEARING	2
23	AN-4-35	BOLT	2
	132100-4-7	BOLT	8
	AN-310-4	NUT	2
	AN 210-4A	PULLEY	2
	AC 365-D428	NUT	8

SECTION **D-D**

OF AIRPLANE

SECTION **A-A**
ROTATED COUNTERCLOCKWISE 60°

SECTION **C-C**

SECTION **B-B**

CONTROL COLUMN

REFER TO DOUGLAS DWG. 5002644

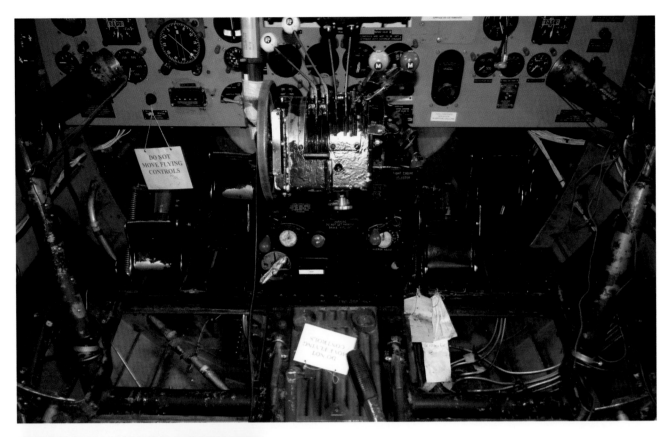

ABOVE Cockpit, with seats and floor removed. Also removed are the pilot's and co-pilot's control wheels to allow for flying control cable replacement. *(Paul Blackah/Crown Copyright)*

LEFT Aileron stripped of its fabric cover. *(Dennis Lovesday)*

BELOW Elevator stripped, showing its construction. *(Dennis Lovesday)*

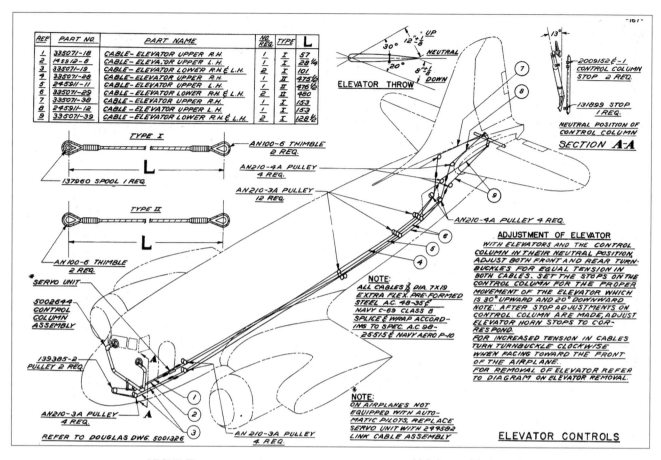

The following table appears within the diagram:

REF	PART NO.	PART NAME	NO. REQ.	TYPE	L
1	335071-18	CABLE-ELEVATOR UPPER R.H.	1	I	57
2	145912-8	CABLE-ELEVATOR UPPER L.H.	1	I	28 1/4
3	335071-19	CABLE-ELEVATOR LOWER R.H. & L.H.	2	I	101
4	335071-28	CABLE-ELEVATOR UPPER R.H.	1	II	475 1/2
5	245911-11	CABLE-ELEVATOR UPPER R.H.	1	II	476 1/2
6	335071-29	CABLE-ELEVATOR LOWER R.H. & L.H.	2	II	480
7	335071-38	CABLE-ELEVATOR UPPER R.H.	1	I	153
8	245911-12	CABLE-ELEVATOR UPPER L.H.	1	I	153
9	335071-39	CABLE-ELEVATOR LOWER R.H. & L.H.	2	I	128 1/2

TYPE I

AN100-6 THIMBLE 2 REQ.

137960 SPOOL 1 REQ.

TYPE II

AN100-6 THIMBLE 2 REQ.

* SERVO UNIT

5002644 CONTROL COLUMN ASSEMBLY

139385-2 PULLEY 2 REQ.

AN210-3A PULLEY 4 REQ.

REFER TO DOUGLAS DWG. 5001326

AN210-4A PULLEY 4 REQ.

AN210-3A PULLEY 12 REQ.

AN210-4A PULLEY 4 REQ.

NOTE: ALL CABLES 1/8 DIA. 7X19 EXTRA FLEX. PRE-FORMED STEEL A.C. 48-35E & NAVY C-69 CLASS B SPLICE & WRAP ACCORD-ING TO SPEC. A.C. 98-25515 & NAVY AERO P-10

* NOTE: ON AIRPLANES NOT EQUIPPED WITH AUTO-MATIC PILOTS, REPLACE SERVO UNIT WITH 249582 LINK CABLE ASSEMBLY

AN210-3A PULLEY 4 REQ.

ELEVATOR THROW — UP 12° +1/3, NEUTRAL, 30°, 20°, 8° +1/2 DOWN

2009152 & -1 CONTROL COLUMN STOP 2 REQ.

131899 STOP 1 REQ.

13°

NEUTRAL POSITION OF CONTROL COLUMN

SECTION A-A

ADJUSTMENT OF ELEVATOR

WITH ELEVATORS AND THE CONTROL COLUMN IN THEIR NEUTRAL POSITION, ADJUST BOTH FRONT AND REAR TURN-BUCKLES FOR EQUAL TENSION IN BOTH CABLES. SET THE STOPS ON THE CONTROL COLUMN FOR THE PROPER MOVEMENT OF THE ELEVATOR WHICH IS 30° UPWARD AND 20° DOWNWARD. NOTE: AFTER STOP ADJUSTMENTS ON CONTROL COLUMN ARE MADE, ADJUST ELEVATOR HORN STOPS TO COR-RESPOND. FOR INCREASED TENSION IN CABLES TURN TURNBUCKLE CLOCKWISE WHEN FACING TOWARD THE FRONT OF THE AIRPLANE. FOR REMOVAL OF ELEVATOR REFER TO DIAGRAM ON ELEVATOR REMOVAL.

ELEVATOR CONTROLS

ABOVE Elevator control diagram. *(AN-01-40NC-2)*

BELOW Elevator control bell crank and operating tube. The elevator connects to the flat plate at the end of the tube using four bolts. *(Paul Blackah/Crown Copyright)*

which is provided to offset part of the weight of the elevator and to reduce the possibility of 'flutter' in flight.

Elevator

Elevators are moved by fore and aft movement of the control column. The forward movement is limited at the lower end of the control column horn by an adjustable stop attached to the fuselage structure. The rearward movement is limited by adjustable bolts located at the bend in the upper elbows of the control columns. These bolts strike against fixed stops mounted on the fuselage structure. A torque tube connects the control columns, and the elevator operating cable assembly attaches to a single horn, mounted at the centre of the torque tube, beneath the pilot's floor.

The cables continue aft through a series of pulleys and fair leads and connect to a single horn, mounted on a torque tube, which interconnects the two elevators. Fitted on the tail structure, forward of the horn, is a stop bracket incorporating rubber stops to limit the travel of the elevators in either direction.

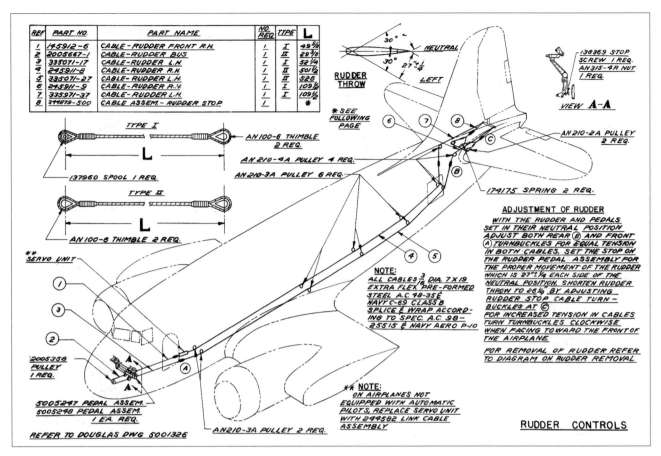

Rudder

Rudder movement is controlled by the rudder pedals, which are adjustable according to the size of the pilots. The two left rudder pedals are connected by a torque tube, which incorporates a single horn. The right pedals system mirrors the left. Adjustable bolts on the pedal torque tube horns, striking against stops on the fuselage structure, limit movement of the rudder pedals. Cables extend from the horns, on the torque tubes, aft through pulleys and fair leads, to the tail section, where they pass up and aft again to connect to the rudder torque tube. A stop cable assembly is forward of, and

ABOVE Rudder control diagram. *(AN-01-40NC-02)*

FAR LEFT Fuselage nose section with access panel removed, giving access to rudder and elevator control cables. *(Paul Blackah/Crown Copyright)*

LEFT Rear fuselage; inside is the rubber torque tube and bell crank. *(Paul Blackah/Crown Copyright)*

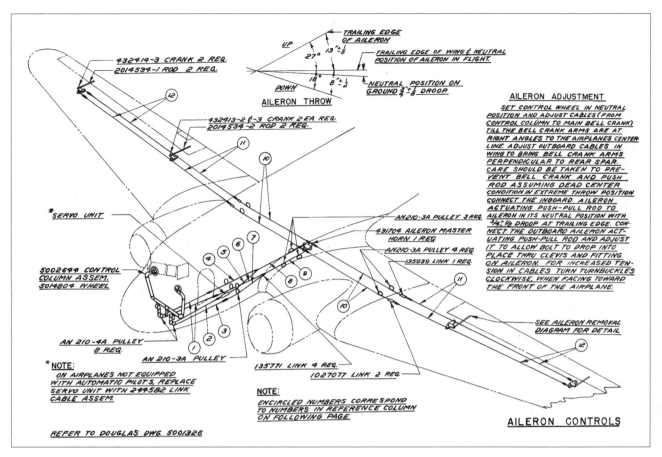

TRAILING EDGE OF AILERON

UP

27° 13"±⅛

TRAILING EDGE OF WING & NEUTRAL POSITION OF AILERON IN FLIGHT.

18° 8"±½

DOWN

AILERON THROW

NEUTRAL POSITION ON GROUND ¾"±½ DROOP

432413-2 & -3 CRANK 2 EA REQ.
2014534-2 ROD 2 REQ.

*SERVO UNIT

5002644 CONTROL COLUMN ASSEM.
5014804 WHEEL

AN 210-4A PULLEY 8 REQ.

AN 210-3A PULLEY

*NOTE:
ON AIRPLANES NOT EQUIPPED WITH AUTOMATIC PILOTS, REPLACE SERVO UNIT WITH 244582 LINK CABLE ASSEM.

REFER TO DOUGLAS DWG. 5001326

AN210-3A PULLEY 3 REQ.
431704 AILERON MASTER HORN 1 REQ.
AN210-3A PULLEY 4 REQ.
135039 LINK 1 REQ.

135771 LINK 4 REQ.
1027077 LINK 2 REQ.

NOTE:
ENCIRCLED NUMBERS CORRESPOND TO NUMBERS IN REFERENCE COLUMN ON FOLLOWING PAGE.

AILERON ADJUSTMENT

SET CONTROL WHEEL IN NEUTRAL POSITION AND ADJUST CABLES (FROM CONTROL COLUMN TO MAIN BELL CRANK) TILL THE BELL CRANK ARMS ARE AT RIGHT ANGLES TO THE AIRPLANES CENTER LINE. ADJUST OUTBOARD CABLES IN WING TO BRING BELL CRANK ARMS PERPENDICULAR TO REAR SPAR. CARE SHOULD BE TAKEN TO PREVENT BELL CRANK AND PUSH ROD ASSUMING DEAD CENTER CONDITION IN EXTREME THROW POSITION CONNECT THE INBOARD AILERON ACTUATING PUSH-PULL ROD TO AILERON IN ITS NEUTRAL POSITION WITH ¾"±½ DROOP AT TRAILING EDGE. CONNECT THE OUTBOARD AILERON ACTUATING PUSH-PULL ROD AND ADJUST IT TO ALLOW BOLT TO DROP INTO PLACE THRU CLEVIS AND FITTING ON AILERON. FOR INCREASED TENSION IN CABLES TURN TURNBUCKLES CLOCKWISE, WHEN FACING TOWARD THE FRONT OF THE AIRPLANE.

SEE AILERON REMOVAL DIAGRAM FOR DETAIL

AILERON CONTROLS

ABOVE Aileron control diagram. (AN-01-40NC-2)

OPPOSITE TOP Elevator tab control diagram. (AN-01-40NC-2)

RIGHT Pilot's centre console below the engine control levers, showing the aileron rudder trim handles and indication dials, fuel tank selectors for the left and right engines, parking brake, tail lock and auto-pilot selector. (Paul Blackah/ Crown Copyright)

OPPOSITE BOTTOM Rudder trim control diagram. (AN-01-40NC-2)

connected to, the rudder horn in order to limit travel of the rudder in either direction.

Ailerons

Aileron movement is controlled by rotation of the wheels on the control columns. Wheel movement and aileron travel is limited by stops on the chain in the control column head. Control cables connect to the chain, which passes over a sprocket on the shaft of each control wheel in the control column. The cables then continue downwards through the centre of each control column and out of the bottom,

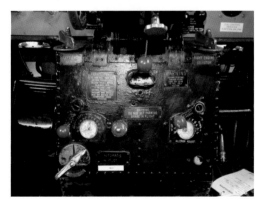

where they extend aft to the centre wing, connecting to two of the four arms of the main operating bell crank. From the other two arms, cables connect to three arm bell cranks, located forward of the ailerons. The control wheel inputs fore and aft movement to the rods, causing up and down movement to the ailerons.

Elevator trim

A hand wheel, installed on the left side of the pilot's control pedestal, controls the elevator trim tabs. Control cables are wound around a drum within the pedestal and extend aft to the tail section, where they diverge into each elevator and wind around another drum located within the elevator itself. Turning the control wheel causes the drums to rotate; forward rotation corrects tail-heavy conditions and aft rotation corrects nose-heavy conditions. A graduated dial, adjacent to the control wheel indicates the setting of the tab.

Rudder trim

The control for the rudder tab is located to the lower left side of the front face of the engine

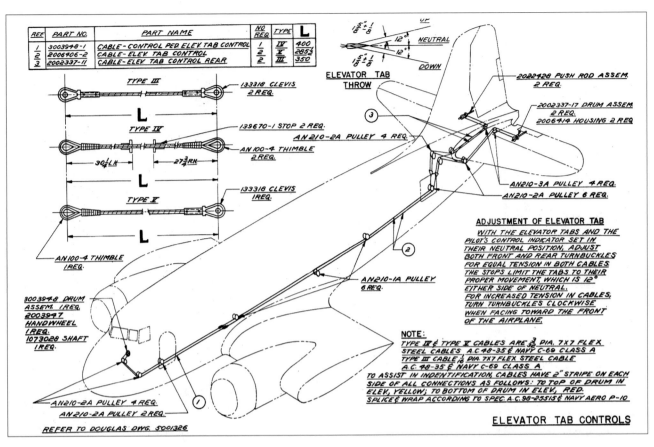

REF	PART NO.	PART NAME	NO. REQ	TYPE	L
1	3003948-1	CABLE-CONTROL PED ELEV TAB CONTROL	1/2	IV	400
2	2006406-2	CABLE-ELEV TAB CONTROL	2	IV	2851
3	2002337-11	CABLE-ELEV TAB CONTROL REAR	2	III	350

TYPE III
133318 CLEVIS 2 REQ.
139670-1 STOP 2 REQ.

TYPE IV
$30\frac{1}{4}$ LH $27\frac{3}{4}$ RH
AN100-4 THIMBLE 2 REQ.

TYPE V
133318 CLEVIS 1 REQ.

AN100-4 THIMBLE 1 REQ.

3003948 DRUM ASSEM. 1 REQ.
2003947 HANDWHEEL 1 REQ.
1073028 SHAFT 1 REQ.

AN210-2A PULLEY 4 REQ.
AN210-2A PULLEY 2 REQ.

REFER TO DOUGLAS DWG. 5001326

ELEVATOR TAB THROW
$\frac{5}{8}^{"} \pm \frac{1}{8}$ UP
12°   NEUTRAL
12°
$1\frac{5}{8}^{"} \pm \frac{1}{8}$ DOWN

2022428 PUSH ROD ASSEM. 2 REQ.
2002337-17 DRUM ASSEM. 2 REQ.
2006414 HOUSING 2 REQ.
AN210-2A PULLEY 4 REQ.
AN210-3A PULLEY 4 REQ.
AN210-2A PULLEY 6 REQ.
AN210-1A PULLEY 6 REQ.

ADJUSTMENT OF ELEVATOR TAB
WITH THE ELEVATOR TABS AND THE PILOT'S CONTROL INDICATOR SET IN THEIR NEUTRAL POSITION, ADJUST BOTH FRONT AND REAR TURNBUCKLES FOR EQUAL TENSION IN BOTH CABLES. THE STOPS LIMIT THE TABS TO THEIR PROPER MOVEMENT, WHICH IS 12° EITHER SIDE OF NEUTRAL. FOR INCREASED TENSION IN CABLES, TURN TURNBUCKLES CLOCKWISE WHEN FACING TOWARD THE FRONT OF THE AIRPLANE.

NOTE:
TYPE IV & TYPE V CABLES ARE $\frac{3}{32}$ DIA. 7X7 FLEX STEEL CABLES A.C. 48-35 & NAVY C-69 CLASS A. TYPE III CABLE $\frac{1}{16}$ DIA. 7X7 FLEX STEEL CABLE A.C. 48-35 & NAVY C-69 CLASS A. TO ASSIST IN INDENTIFICATION, CABLES HAVE 2" STRIPE ON EACH SIDE OF ALL CONNECTIONS AS FOLLOWS: TO TOP OF DRUM IN ELEV, YELLOW; TO BOTTOM OF DRUM IN ELEV, RED. SPLICE & WRAP ACCORDING TO SPEC. A.C. 98-23515 & NAVY AERO P-10

ELEVATOR TAB CONTROLS

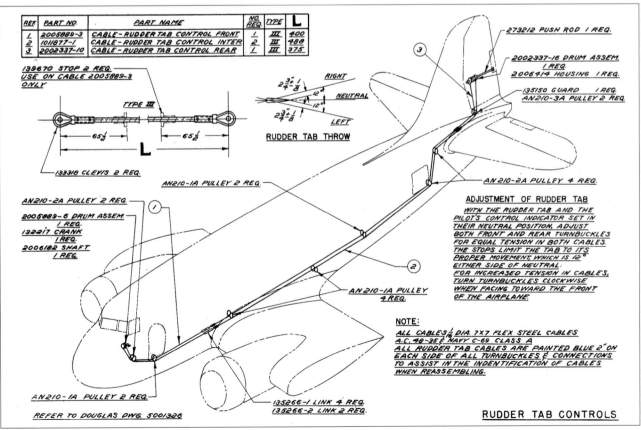

REF	PART NO.	PART NAME	NO. REQ	TYPE	L
1	2005889-3	CABLE-RUDDER TAB CONTROL FRONT	1	III	400
2	101877-1	CABLE-RUDDER TAB CONTROL INTER	2	III	488
3	2002337-10	CABLE-RUDDER TAB CONTROL REAR	1	III	375

139670 STOP 2 REQ.
USE ON CABLE 2005889-3 ONLY

TYPE III
$65\frac{1}{2}$ $65\frac{1}{2}$
133318 CLEVIS 2 REQ.

AN210-2A PULLEY 2 REQ.
2005889-6 DRUM ASSEM. 1 REQ.
132217 CRANK 1 REQ.
2006182 SHAFT 1 REQ.

AN210-1A PULLEY 2 REQ.

RUDDER TAB THROW
$2\frac{3}{4}^{"} \pm \frac{1}{8}$ RIGHT
12°   NEUTRAL
12°
$2\frac{3}{4}^{"} \pm \frac{1}{8}$ LEFT

273212 PUSH ROD 1 REQ.
2002337-16 DRUM ASSEM. 1 REQ.
2006414 HOUSING 1 REQ.
135150 GUARD 1 REQ.
AN210-3A PULLEY 2 REQ.
AN210-2A PULLEY 4 REQ.
AN210-1A PULLEY 2 REQ.
AN210-1A PULLEY 4 REQ.

ADJUSTMENT OF RUDDER TAB
WITH THE RUDDER TAB AND THE PILOT'S CONTROL INDICATOR SET IN THEIR NEUTRAL POSITION, ADJUST BOTH FRONT AND REAR TURNBUCKLES FOR EQUAL TENSION IN BOTH CABLES. THE STOPS LIMIT THE TAB TO ITS PROPER MOVEMENT, WHICH IS 12° EITHER SIDE OF NEUTRAL. FOR INCREASED TENSION IN CABLES, TURN TURNBUCKLES CLOCKWISE WHEN FACING TOWARD THE FRONT OF THE AIRPLANE.

NOTE:
ALL CABLES $\frac{1}{16}$ DIA. 7X7 FLEX STEEL CABLES A.C. 48-35 & NAVY C-69 CLASS A. ALL RUDDER TAB CABLES ARE PAINTED BLUE 2" ON EACH SIDE OF ALL TURNBUCKLES & CONNECTIONS TO ASSIST IN THE INDENTIFICATION OF CABLES WHEN REASSEMBLING.

135266-1 LINK 4 REQ.
135266-2 LINK 2 REQ.

REFER TO DOUGLAS DWG. 5001326

RUDDER TAB CONTROLS

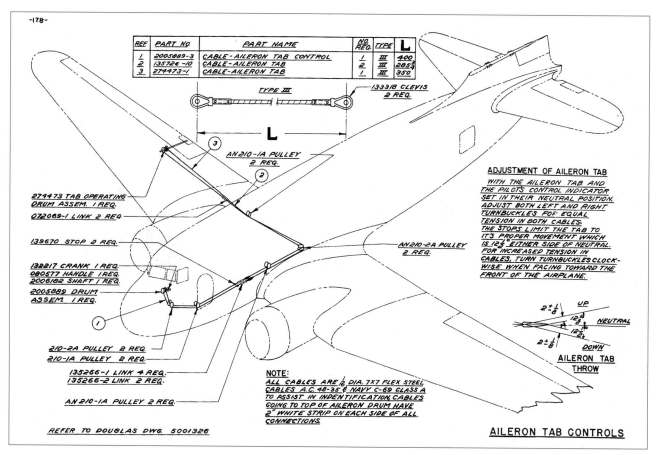

REF	PART NO.	PART NAME	NO. REQ.	TYPE	L
1	2005889-3	CABLE-AILERON TAB CONTROL	1	III	400
2	135726-10	CABLE-AILERON TAB	2	III	285¾
3	274473-1	CABLE-AILERON TAB	1	III	350

TYPE III

-133318 CLEVIS
2 REQ.

L

3
2
AN 210-1A PULLEY
2 REQ.

274473 TAB OPERATING
DRUM ASSEM. 1 REQ.

072069-1 LINK 2 REQ.

139670 STOP 2 REQ.

132217 CRANK 1 REQ.
080577 HANDLE 1 REQ.
2006182 SHAFT 1 REQ.
2005889 DRUM
ASSEM. 1 REQ.

1

210-2A PULLEY 2 REQ.
210-1A PULLEY 2 REQ.

135266-1 LINK 4 REQ.
135266-2 LINK 2 REQ.

AN 210-1A PULLEY 2 REQ.

AN 210-2A PULLEY
2 REQ.

ADJUSTMENT OF AILERON TAB
WITH THE AILERON TAB AND
THE PILOTS CONTROL INDICATOR
SET IN THEIR NEUTRAL POSITION,
ADJUST BOTH LEFT AND RIGHT
TURNBUCKLES FOR EQUAL
TENSION IN BOTH CABLES.
THE STOPS LIMIT THE TAB TO
IT'S PROPER MOVEMENT WHICH
IS 12½° EITHER SIDE OF NEUTRAL.
FOR INCREASED TENSION IN
CABLES, TURN TURNBUCKLES CLOCK-
WISE WHEN FACING TOWARD THE
FRONT OF THE AIRPLANE.

UP
NEUTRAL
DOWN
AILERON TAB
THROW

NOTE:
ALL CABLES ARE 1/16 DIA. 7X7 FLEX STEEL
CABLES A.C. 48-35 & NAVY C-69 CLASS A
TO ASSIST IN INDENTIFICATION, CABLES
GOING TO TOP OF AILERON DRUM HAVE
2" WHITE STRIP ON EACH SIDE OF ALL
CONNECTIONS.

REFER TO DOUGLAS DWG. 5001326

AILERON TAB CONTROLS

ABOVE Aileron trim control diagram. *(AN-01-40NC-2)*

control pedestal. Within the pedestal, the tab control cables are wound round a drum, similar to the elevator trim. The cables then pass over a set of pulleys below the pilot's compartment, aft to the tail section of the aircraft, where they go up into the rudder and wind around a drum. This unit is mounted on the main torque tube just below the lower rudder hinge. Clockwise rotation of the wheel corrects nose left yaw and anti-clockwise corrects yawing to the right. A graduated dial, adjacent to the control, indicates the setting of the tab.

Aileron trim

Only the right-hand aileron is equipped with a trim tab, which is operated by a lever on the lower right-hand side of the front face of the control pedestal. Tab control cables are wound around a drum within the pedestal and pass over a set of pulleys beneath the pilot's floor. The cables then continue aft of the centre wing section and outboard to a drum located within the aileron. Clockwise rotation of the control corrects a low left-wing flying condition and anti-clockwise rotation corrects a low right-wing flying condition. Again, there is a graduated dial, next to the control, to provide an indication of the setting of the tab.

Ranges of movement of control surfaces

Ailerons	UP 13in +/-0.125in (27°)
	DOWN 8in +/-0.5in (18°)
Aileron droop	1.5°
Aileron trim	UP travel 2in +/-0.125in (12.5°)
	DOWN travel 2in +/-0.125in (12.5°)
Elevators	UP 12in +/-0.125in (30°)
	DOWN 8in +/-0.125in (20°)
Elevator trim tabs	UP 1.625in +/-0.125in (12°)
	DOWN 1.625in +/-0.125in (12°)
Rudder	RIGHT 27.125in +/-0.125in (30°)
	LEFT 27.125in +/-0.125in (30°)
Rudder trim	RIGHT 2.75in +/-0.125in (12°)
	LEFT 2.75in +/-0.125in (12°)
Flaps	FULLY DOWN 45°

Wing flaps

The wing flaps are of split trailing edge type, extending between the inboard edges of the ailerons. The flaps comprise four sections, which function as a single continuous unit. They are attached to the underside of the wing by piano hinges. The flaps are hydraulically operated and a hydraulic jack is located in the centre wing panel. The jack consists of a piston and a longitudinally acting cylinder, mounted on rollers. Toggle rods are connected to both the cylinder and the piston, forming a variable parallelogram, giving equalised movement. Push-pull rods, connected to the exterior point of the toggle rods, are attached to turnbuckles on the wing flaps. Lengthening of the hydraulic jack causes the push-pull rods to move inwards, lowering the flaps; the reverse action causes the wing flaps to rise.

For operation and control of the flaps, see 'Hydraulic system'.

LEFT A view of the flap jack and the rail that it travels on when the flaps are raised and lowered. *(Paul Blackah/Crown Copyright)*

LEFT Flap operating linkages. These are connected to a tube, which moves outboard or inboard depending on whether the flaps are raising or lowering. *(Paul Blackah/Crown Copyright)*

Fire protection system

The fire protection system consists of two fire bottles, located in each engine nacelle in the undercarriage bay. Four 'flame' switches, located in the cockpit, electrically operate these fire bottles. Fire wire, attached to the engine, will give an indication of a fire by illuminating a warning light on the instrument panel. The pilot will then push the relevant 'flame' switch, which will discharge the bottle through a spray ring attached around the engine, putting the fire out.

One bottle is discharged on the engine and the other bottle is discharged around the carburettor. If the aircraft crashes, such as a wheels-up landing, a crash switch will automatically set off all the bottles.

Radio and navigational aids

Aircraft that are flying today, whether modern or historic, are obliged to carry modern radios and navigational equipment in order to conform with current flying regulations.

The BBMF Dakota is fitted with two V/UHF

BELOW LEFT Electrically operated fire bottles for the engine. *(Paul Blackah/ Crown Copyright)*

BELOW The engine fire extinguisher system. *(AN-01-40NC-2)*

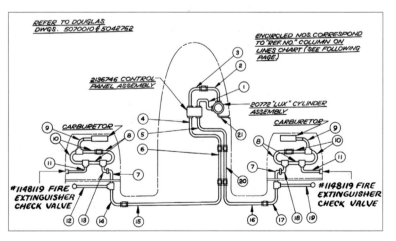

radios: a PTR1751 for UHF and an AD120 for VHF. The aircraft is also fitted with an Instrument Landing System (ILS), a VHF Omni-directional Ranging system, a Distance Measuring positioning system (DME) and an Identification Friend or Foe system. All the equipment is located in the cockpit and on the radio rack opposite the navigator's table.

There is also an intercom system, consisting of three crew boxes and eight plug-in points in the fuselage. This allows the crew to be in communication with the ground crew and each other.

Instrument panel

The main instrument panel is manufactured in sections, which are bolted together to form a single unit. The panel is attached to the aircraft structure by means of five shock-absorber mounts. The panel is illuminated by six lights: three fluorescent and three incandescent.

Pitot static instruments and equipment consist of two airspeed indicators, two altimeters and one rate-of-climb indicator. These are installed on the main instrument panel. Two pitot static tubes are mounted on masts on the underside of the fuselage beneath the pilot's compartment. These tubes are protected from ice by a built-in electrical heater.

The vacuum-operated instruments and equipment consist of a gyro-horizon indicator, a turn and bank indicator and the automatic pilot gyro control unit. These are operated by two engine-mounted vacuum pumps.

Pilot's compass

The BBMF Dakota is fitted with a G4B gyro-magnetic compass, mounted in the instrument panel on the pilot's side, with a master indicator located above the navigator's table and a detector unit mounted in the wing. An emergency E2B compass is fitted in the 'V' of the windscreens. This is used in the event of failure of the main system.

Transport and cargo fits

The main cabin has a variety of uses. Many Dakotas have passenger seats fitted in this section; however, a wartime Dakota and the

ABOVE Radio control panel, below which are two red buttons for feathering the propellers. In between the red buttons are the engine's magneto switches and below them is the E2B standby compass. At the top, the two red handles unlock the pilot compartment emergency escape hatch. *(Paul Blackah/ Crown Copyright)*

RIGHT Radio rack on the starboard side, aft of the pilot's compartment. *(Paul Blackah/Crown Copyright)*

RIGHT Inside the nose bonnet, looking at the rear of the instrument panel. *(Paul Blackah/ Crown Copyright)*

BBMF Dakota were fitted with folding bench type seats along each side of the main cabin compartment, allowing seating for 28 personnel. These seats can be folded down to allow more room. This 'fit' allows for para-dropping; a static line is fitted from the front main cabin bulkhead to the third frame forward of the aft bulkhead. A jump signal light box, fitted with a red lamp and a green lamp, is located forward of the cargo loading door. When the light box shows red, the paras prepare to jump. When the light turns green they jump. They are clipped to the static line and on leaving the aircraft their parachutes are automatically deployed.

The aircraft was also used in the transportation of wounded personnel. There are two types of fit for this purpose: the metal type and the strap type. The metal type allows for 18 stretchers fitted onto curved metal supports. The strap type allows for 24 stretchers suspended on canvas straps fitted to the fuselage sides.

Another use of the aircraft was to carry large loads, such as four Pratt & Whitney engines, in their cradles, three Allison V-1710 engines, airborne task force equipment, consisting of a ¼ton jeep, and one 37mm anti-tank gun, mounted on its carriage, or a 75mm field Howitzer. Some Dakotas were fitted out to carry mules, which would offer the ground troops a means of transportation over difficult terrain.

On the underside of the aircraft, six parachute-pack containers could be fitted. It was also possible to fit a four-bladed and three-bladed propeller making full use of the aircraft's capacity.

Some Dakotas were modified to tow gliders and this was done by fitting a release unit, which was attached by four tube assemblies, bolted to the fuselage structure within the tail cone. Two heavy 'jaws' inside the release unit engage a ring in the glider tow attachment. A cable-operated handle marked 'GLIDER RELEASE', which is located directly at the back of the co-pilot's seat, opens these jaws.

Many of the senior ranks had their own Dakotas for personal use. This gave ease of transportation with the added bonus of enabling work to be carried out while on the move. The interior would be fitted out and contain items according to each person's choice.

ABOVE **Passenger seat fit on G-AMPY.** (Tim Badham)

LEFT **Red and green para jump lights near the rear door.** (Paul Blackah/Crown Copyright)

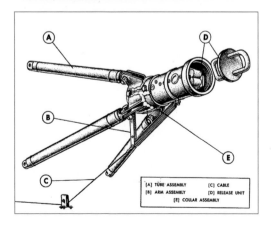

LEFT **The glider tow assembly.** (AN-01-40NC-2)

RIGHT Cockpit

1 Airspeed indicator
2 Altimeter
3 Gyro unit (GHB compass)
4 Vacuum gauge
5 Flap indicator
6 Press to transmit/intercom switch
7 Artificial horizon
8 Windscreen wiper motors
9 Radio magnetic indicator
10 Localiser glide slope indicator
11 Turn and slip indicator
12 Clock/stopwatch
13 Manifold pressure selector
14 rpm
15 Manifold pressure
16 Oil pressure
17 Fuel pressure
18 Cylinder head temperature
19 Oil temperature
20 Carburettor temperature
21 Fuel contents
22 Fire extinguisher/warning
23 Windscreen de-icer
24 Autopilot pressure
25 Alternate static selector
26 Cabin outer heat indicator
27 Undercarriage DOWN lights, green
28 Undercarriage UP lights, red
29 Press to transmit/intercom switch
30 Marker lights (ILS)
31 Auto-pilot heading
32 Auto-pilot attitude
33 Auto-pilot vacuum
34 Fire wire test
35 Fire wire test
36 Hydraulic pressure gauges
37 Port fuel tank selector
38 Starboard fuel tank selector
39 Rudder trim handle
40 Parking brake
41 Aileron trim handle
42 Pitch trim wheel
43 Pitch control levers
44 Throttle control levers
45 Mixture control levers
46 Mixture control levers
47 Carburettor air intake
48 Pilot's rudder pedals
49 Co-pilot's rudder pedals
50 Seat adjusting cable
51 Hydraulic hand pump
52 Dog latch
53 Friction damper throttle
54 Tail-wheel lock

Chapter Three

The Pratt & Whitney R-1830-90D engine

The original DC-1 prototype aircraft was fitted with a Wright Cyclone 9-cylinder engine, but as the aircraft design evolved the engine of choice became the Pratt & Whitney R-1830 series. These rugged engines have become as reliable as the Dakota itself and the majority of Dakotas flying today are still fitted with them, although the technology is now available to modify the aircraft to take Pratt & Whitney turboprop engines. For the purpose of this chapter we will be taking a closer look at the original Pratt & Whitney R-1830-90D engine, as fitted on the BBMF Dakota.

OPPOSITE Engine mounted on port nacelle with prop removed. *(Jim Douthwaite/Crown Copyright)*

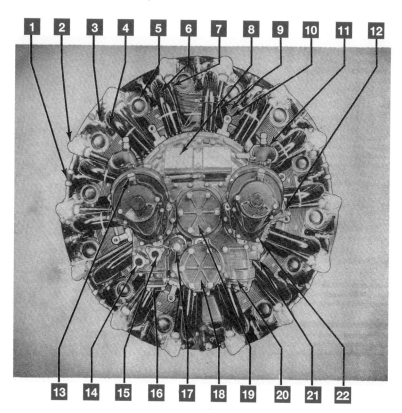

LEFT Rear view of the engine diagram.
(AN-01-40NC-2)

1 Rocker box inter-cylinder oil drainpipe
2 Valve rocker box
3 Spark plug lead elbow
4 Spark plug
5 Cylinder exhaust port
6 Intake pipe
7 Push rod cover tube
8 Carburettor mounting pad
9 Fuel primer lines
10 Engine mount attachment bracket
11 Ignition conduit
12 Auxiliary drive gear mounting pad
13 Magneto ground connection
14 Oil inlet port
15 Oil outlet port
16 Fuel pump drive
17 Oil-pressure relief valve and temperature compensating piston
18 Generator mounting pad
19 Starter mounting pad
20 Oil strainer chamber
21 Oil strainer bypass valve
22 Magneto

The engine is a 14-cylinder, twin-row and radial air-cooled, with a 16:9 propeller reduction gear ratio, and is fitted with a Hamilton Standard Hydromatic full-feathering propeller.

LEFT Front view of the engine diagram.
(AN-01-40NC-2)

1 Valve rocker box
2 Push rod cover tube
3 Rocker box oil drainpipe
4 Spark plug ignition conduits
5 Ignition harness
6 Propeller governor mounting pad
7 Engine nose case

Crankshaft

The crankshaft is a solid, one-piece steel forging, supported by large roller bearings located at each end. The cylinders and crankshaft bearings are held in place by a three-section crankcase of forged aluminium alloy to give it strength.

The engine is designed to be free from any critical vibration by virtue of specially designed dynamic balancing devices installed in the crankshaft. The master rods are steel forgings of two-piece construction and are fitted with special silver bearings capable of withstanding high bearing loads for long periods.

Cylinders

The cylinders are constructed of cast-aluminium alloy heads, shrunk tightly onto steel barrels, having a large number of deep, closely spaced integral cooling fins. All cylinders are fitted with specially designed baffles, which force the cooling air to pass between the fins, so as to obtain the most efficient cooling possible. The intake and exhaust valve mechanisms, such as the rocker arms, valve springs and valves, are contained in housings, which are an integral part of the cylinder head. Constant lubrication of these parts is obtained by force-feed of engine oil under pressure.

ABOVE Pratt & Whitney R-1830 crankshaft and con rods *(Paul Blackah/ Crown Copyright)*

LEFT The crankshaft bearing has seized and the crankshaft is badly scored where the engine has overheated and has turned the metal blue. *(Paul Blackah/Crown Copyright)*

FAR LEFT Champion spark plug used on the engine. There are 28 spark plugs per engine. *(Paul Blackah/ Crown Copyright)*

LEFT Engine manufacturer's data plate mounted on the reduction gear housing. *(Paul Blackah/ Crown Copyright)*

RIGHT Reduction gear
and propshaft. The
chrome ring carries
the wiring harness
to the spark plugs.
*(Paul Blackah/Crown
Copyright)*

Pistons

The pistons are machined from aluminium alloy forgings and are held within close tolerances of weight to ensure proper balancing. Each piston incorporates five piston rings: three compression rings in the three top grooves, the fourth groove holds a dual oil ring and the fifth holds an oil scraper ring. Recesses in the head of the piston allow for clearance of the inlet and exhaust valves.

Rocker arms and boxes

The two rocker housings on the cylinder head each contain two concentric valve springs and a valve rocker arm. Each pair of valve springs is secured to a valve stem by a split cone and washer. The exhaust and inlet valve springs are interchangeable and may be withdrawn from the housing without removing the rocker arms. Safety circlets are fitted on the

BELOW Piston ring
removal diagram.
(AN-01-40NC-2)

BELOW RIGHT Piston
ring arrangement
chart. *(AN-01-40NC-2)*

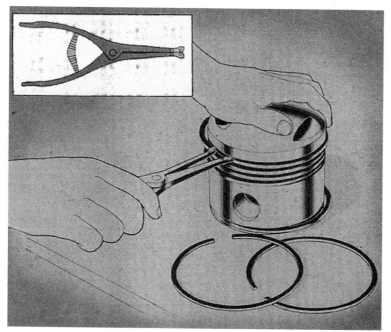

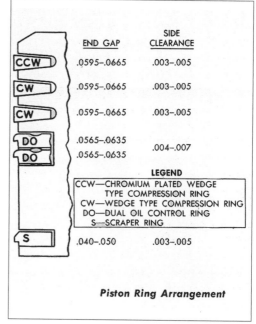

	END GAP	SIDE CLEARANCE
CCW	.0595–.0665	.003–.005
CW	.0595–.0665	.003–.005
CW	.0595–.0665	.003–.005
DO	.0565–.0635	.004–.007
DO	.0565–.0635	
S	.040–.050	.003–.005

LEGEND
CCW—CHROMIUM PLATED WEDGE
 TYPE COMPRESSION RING
CW—WEDGE TYPE COMPRESSION RING
DO—DUAL OIL CONTROL RING
S—SCRAPER RING

Piston Ring Arrangement

valve stems to avoid the possibility of the valves dropping into the cylinders when the split cones are removed. Each rocker arm is supported on a shaft by a double-row ball bearing and has a valve clearance adjustment screw. Drilled passages give each rocker arm direct-pressure oil to the ball bearings and valve adjusting screw. Studs and lock nuts to the top of the rocker housing secure rocker box covers and oil drainpipes connect the two rocker boxes of each cylinder.

Valves

Each cylinder contains an inlet and exhaust valve. A steel insert in the valve adjusting screw (installed in the rocker arm) is used as a contact between the adjusting screw and the valve stem to minimise the friction at this point. Exhaust valves have hollow heads and stems and are sodium filled.

Supercharger

The supercharger is an integral, single-speed unit, constructed of an aluminium housing containing a large, dynamically balanced impeller, mounted on a steel shaft, which is supported by ball bearings at each end. This shaft is driven at 7.15 times the crankshaft speed. The supercharger section is also fitted with eight forged steel pedestals, through which the engine is bolted to the engine mount.

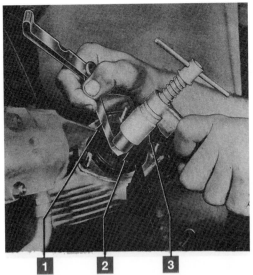

LEFT Adjusting valve clearances diagram.
(AN-01-40NC-2)

1 Valve clearance feeler gauge
2 Rocker arm
3 Valve clearance adjustment tool

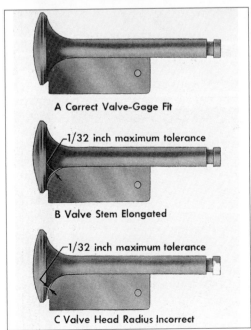

LEFT Diagram showing the rocker arm installation.
(AN-01-40NC-2)

1 Rocker box
2 Valve spring
3 Valve and split lock
4 Valve clearance adjustment screw
5 Rocker arm
6 Rocker arm shaft and shaft nut

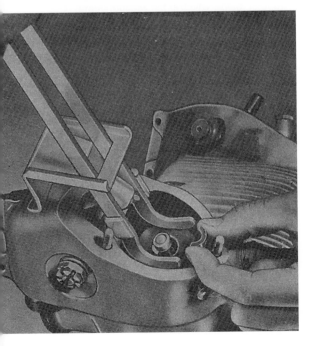

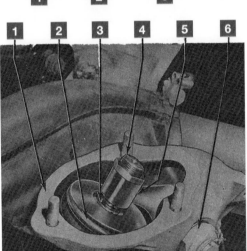

A Correct Valve-Gage Fit

—1/32 inch maximum tolerance

B Valve Stem Elongated

—1/32 inch maximum tolerance

C Valve Head Radius Incorrect

FAR LEFT Installing a split lock on the exhaust valve stem diagram.
(AN-01-40NC-2)

LEFT Valve radius stretch gauge diagram.
(AN-01-40NC-2)

Ancillary section

This section of the engine is aluminium alloy casting fitted with a number of mounting pads, which are provided with drives for the various accessories, such as hydraulic pump, oil pump, fuel pump, generator, starter motor and vacuum pump. Two magnetos are bolted onto special mounting pads, the left magneto fires the rear spark plugs and the right magneto fires the front spark plugs.

LEFT Rear of the engine, showing the ancillary drives for the magnetos. The generator, starter and carb attachment points are blanked off. *(Paul Blackah/Crown Copyright)*

LEFT Rear of engine with the magnetos fitted (one with cover removed); the electric starter motor (top) with generator fitted below. *(Paul Blackah/Crown Copyright)*

BELOW LEFT Rear of the engine showing the generator, below which is the vacuum pump. *(Paul Blackah/Crown Copyright)*

BELOW RIGHT A close-up of the installation of one of the engine magnetos. *(Paul Blackah/ Crown Copyright)*

Engine specifications

Model	Twin Wasp
Type	2-row, radial, air-cooled
Number of cylinders	14
Bore	5.5in
Stroke	5.5in
Piston displacement	1,830cu in
Compression ratio	6.7:1
Impeller gear ratio	7.15:1
Diameter of impeller	11in
Crankshaft rotation	clockwise
Propeller rotation	clockwise
Dry weight of engine	1,467lb
Diameter of engine	48.19in
Length of engine	61.67in
Rotation of magneto drive	clockwise
Magneto speed	0.875:1 in multiples of crankshaft speed
Spark plug gap	0.011 to 0.014in
Spark advance	25°

Valves and timing	
Intake opens before top centre	20°
Intake closes after bottom centre	76°
Exhaust opens before bottom centre	76°
Exhaust closes after top centre	20°
Valve adjusting clearance (cold)	0.020in

Engine mount

Each engine mount is a welded assembly of steel tubing, gussets and attachment fittings. When complete, the engine mount is a single rigid structure. It is secured to the firewall on the nacelle of the centre section using four bolts. The mount incorporates eight fittings for the attachment of the engine. These engine mounts are interchangeable to left and right.

ABOVE The rear of the engine mount, showing one of the four attachment bolts that mount it to the aircraft. *(Paul Blackah/ Crown Copyright)*

BELOW LEFT One of eight mounting points on the engine. *(Paul Blackah/Crown Copyright)*

BELOW RIGHT Engine mount and firewall, attached to the power plant stand, waiting for an engine to be fitted. The eight edge rings in the centre of the engine mount are used to secure the engine in place with bolts and rubber mounts. *(Jim Douthwaite/Crown Copyright)*

Exhaust collector ring

This comprises seven sections clamped together; each section is manufactured from corrosion-resistant steel. The complete ring is fastened to the engine attachment fittings on the engine mount at six points, and fits over stacks bolted to the exhaust ports of the engine. A tail pipe, which extends out of the lower outboard side of each engine nacelle, is connected to the exhaust collector ring by an expansion joint.

Engine ring cowling

This is made up of three sections: an upper panel and two lower side panels. The lower side panels are attached to the upper panel with Dzus fasteners and an adjustable fastener in the leading edge. The lower panels are connected at the bottom by adjustable fasteners. The cowling assembly rests on the rocker boxes of the cylinders and is supported by lugs bolted to the rocker boxes of the rear bank of cylinders. The trailing edge of the cowling assembly incorporates hydraulically operated cowl flaps, which are adjustable from the pilot's compartment.

Engine accessory cowlings

These cowlings are installed around the nacelle aft of the engine ring cowling, and are attached by Dzus fasteners. When removed, this gives access to the engine accessories without having to remove the entire cowlings.

Carburettor

The engines are fitted with a Bendix-Stromburg downdraft injection carburettor. The carburettor is made up of five parts: the main throttle body, the automatic mixture control unit, the fuel regulator unit, the fuel control unit and the adaptor unit.

The throttle unit comprises an air scoop, where air flows to a two-barrelled throttle unit. Each barrel contains a large venturi, a boost venturi and eight small impact tubes, which are located around the top of the large venturi. The throttle butterfly valves control the amount of incoming air. The difference in air pressure between the impact tubes and the boost venturi throat is a measure of the amount of air entering the engine.

The automatic mixture control unit is

ABOVE Side view of the engine, showing the arrangement of the exhaust (black) and the fireproof bulkhead rear of the exhaust. The brass pipe in front of the exhaust, going around the engine, is a fire-extinguishing spray ring. *(Paul Blackah/Crown Copyright)*

BELOW The engine exhaust ring assembly is made up of seven separate sections. *(Jim Douthwaite/Crown Copyright)*

mounted in the top of the throttle body and contains a sealed metallic bellows, which is filled with nitrogen and oil and is sensitive to changes in carburettor air density and temperature. This maintains a constant fuel–air ratio, regardless of the density of the air.

The regulator unit has an air chamber and a fuel chamber. Each chamber is divided into sections by diaphragms installed on the main fuel poppet valve shaft. The incoming air pressure produces both an air metering force and a vacuum metering force, which acts on the air diaphragm. This force moves the fuel poppet and the fuel diaphragm, permitting fuel to flow to the main discharge nozzle.

The fuel control unit contains fixed metering jets, the idle mixture metering jet, the power enrichment valve and a four-position mixture selection valve with the following positions: IDLE CUT-OFF, AUTO LEAN, AUTO RICH and EMERGENCY RICH.

The adaptor unit is fitted to the base of the throttle body and acts as the mounting unit to attach the carburettor to the engine. The unit contains the main discharge nozzle and the vacuum-operated acceleration pump. When the throttle butterfly valve is closed, suction on the spring-loaded diaphragm charges the acceleration pump. When the throttle is opened the acceleration pump discharges fuel into the blower throat.

Propeller

A Hamilton Standard hydromatic three metal-bladed propeller is mounted on the propeller shaft of each engine. These propellers are of the quick feathering type. An engine-driven CSU (Constant Speed Unit) is mounted on the front section of each engine. The constant speed of each engine is maintained automatically by its CSU, which operates to cause a change in propeller blade angle, through the constant speed range (1,200rpm minimum to 2,700rpm maximum) to meet changing conditions in altitude, attitude and throttle settings. This change in blade angle is carried out hydraulically by the CSU through regulation of the flow of engine oil under pressure to and from the propeller. Each propeller is equipped with a feathering system, powered by an individual electric motor-driven hydraulic pump, which is

ABOVE Looking down on the power plant, the rectangular section in the middle of the picture is where the carburettor draws its air in. *(Jim Douthwaite/ Crown Copyright)*

BELOW Propeller hub, with prop dome removed; the securing nut is visible in the centre. The two small curved pipes are for the prop de-icing fluid, which is sprayed onto the blades to prevent ice build-up. *(Paul Blackah/Crown Copyright)*

installed on the engine firewall and is electrically operated by the pilot.

The feathering pump supplies oil under high pressure independent of engine oil pressure in order to feather and unfeather the prop blades. An engine would be feathered in flight to test the system or in the event of engine failure. The feathered propeller places the blades at an angle of 88°, which means the blades lie directly in the line of flight, acting as a powerful brake to stop rotation of the engine and, at the same time, reducing drag to a minimum.

Engine controls

Each engine has an individual throttle, carburettor mixture and carburettor hot-

Hamilton Standard propeller dimensions

Basic diameter	11ft 7in
Low angle	16° (measured at the 42in station)
High angle	88° (measured at the 42in station)
Constant speed range	1,200rpm min– 2,700rpm max
Propeller reduction gear ratio	16:9

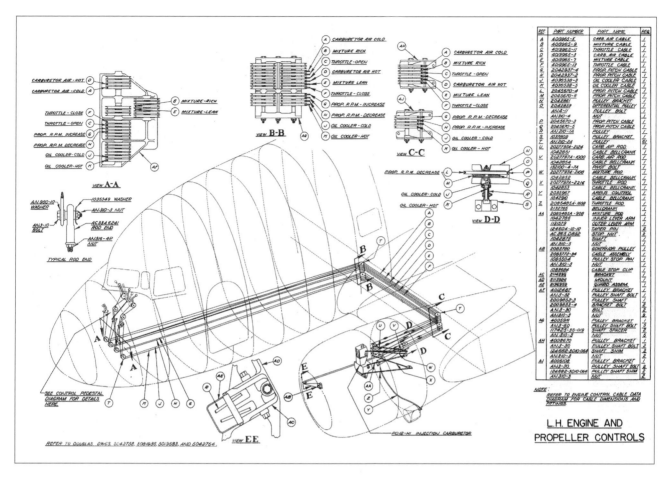

L.H. ENGINE AND PROPELLER CONTROLS

air control. The controls consist of levers located on the control pedestal in the pilot's compartment. Cables connected from the levers, through a bell crank push-pull rod system, operate the controls on the engine. The throttle levers are located on the neck, or quadrant part, of the control pedestal and the knobs are marked with a 'T'. The throttle controls can be locked in any position by applying the lock control. This is a knob located on the underside of the control pedestal, directly under the throttle control levers. Turning the knob clockwise draws the throttle control brake band tight around the throttle control drum and, through friction, locks the levers and prevents creeping of the controls.

The carburettor mixture controls are located next to the throttle levers and are marked with an M. Each of these levers can be secured by a thumb-operated catch-type lock, which enables the control to be locked in either the AUTO RICH or AUTO LEAN position. The spring-loaded catch engages in notches in the quadrant at the above positions. The carburettor

heat controls consist of three levers mounted to the side of the engine control pedestal on the right, co-pilot's side. Two are marked with a C and the lock lever is marked with an L.

Other controls affecting performance of the engines are the propeller pitch controls, the non-ram carburettor filter control and the oil cooler shutter controls.

ABOVE Left-hand engine and propeller control diagram.
(AN-01-40NC-2)

LEFT The engine controls in the cockpit, consisting of a series of levers: two for the throttle, two for the propeller and two for the mixture.
(Paul Blackah/Crown Copyright)

Chapter Four

The Owner's view

The design of the Douglas DC-3 proved so reliable and adaptable that the Dak was adopted by many air forces and airlines the world over and a number are still operating in a variety of roles. The Dakota has a large following, not only among those who have served with the aircraft in one capacity or another, but also among those who have come to appreciate its design qualities. In many parts of the world the DC-3 is still the only means of transportation for goods and people, and the aircraft's longevity is a testament to Donald Douglas's design skills.

OPPOSITE Dakotas N473DC and N1944A taking off in formation. These are both privately owned Dakotas, one based at East Kirkby and the other at Kemble. *(Jarrod Cotter)*

RIGHT Due to prolonged exposure in direct sunlight, the paint has started to blister and peel on the fuselage and propellers. This could lead to corrosion on the skin and prop blades. *(Paul Blackah)*

BELOW The engines on this aircraft would need a complete overhaul as they have not been used for a long while. *(Paul Blackah)*

Purchasing an airworthy Dakota

As there are so many airworthy Dakotas around the world, it is much easier and cheaper to purchase one of these Second World War aircraft than, for example, a fighter like the Spitfire that commands a higher price regardless of whether it is an airworthy example or requires extensive restoration. The only restrictive factor is finding an aircraft that is for sale.

After purchasing your Dakota there is a good support network available, with specialist engine overhaulers in the UK and the USA. Spares are readily obtainable and, in some cases, still being manufactured in the USA.

One of the most important matters to consider is the paperwork that is supplied with the aircraft; it would need its full servicing history to ensure that the components fitted, such as the engines, are not out of 'life' and require replacing. This information would be in

the aircraft and engine logbooks and any ADs (Airworthiness Directives) issued by Douglas (or since the takeover, by Boeing) that are relevant to the aircraft must have been carried out. All this information would make it easier to get your aircraft registered with the CAA.

Purchasing a non-flying restoration project

There are still aircraft available to purchase as an ongoing, non-flying restoration project, with the aim of restoring the aircraft to either a static display or an airworthy condition. This type of project would require careful consideration as, depending on condition, your project could become *very* expensive and it could be cheaper to buy an airworthy aircraft that needs little or no work.

The condition of the aircraft for a project such as this would depend on various factors, including whether it has been stored inside

ABOVE The dusty interior of the main cabin of a Dakota, showing the sort of condition to expect for a possible restoration project. *(Paul Blackah)*

BELOW The rear bulkhead of the main compartment covered in dust and the floor littered with debris. This aircraft would need a thorough paint strip and inspection to assess the viability of restoration. *(Paul Blackah)*

ABOVE RIGHT Inside the undercarriage compartment of the same aircraft, showing the corrosion on the skin, corrosion on the chromed retraction jack ram and compensator jack ram. *(Paul Blackah)*

BELOW RIGHT All rubber hoses have perished, flying control cables are corroded and the bird's nest would also need removing. *(Paul Blackah)*

or outside, and in what type of climate – a damp, humid climate or a dry, hot one. The climate will affect the condition of the aircraft in differing ways. When was the last time the aircraft moved? Tyres and hoses perish easily and engines could be seized. Flying control fabric could be rotted through and the Perspex could have turned milky or cracked, allowing the elements access to the interior. If the aircraft paintwork is not intact then corrosion becomes a major consideration.

Insurance and running costs

The cost of insuring a Dakota depends on each individual case and is calculated according to pilot experience, hours flown per year and what the aircraft is used for. The BBMF's Dakota is a military-owned and operated aircraft and is covered by the MOD.

Civilian owner/operators

There are still numerous civilian and military owner/operators across the world. The Dakota can be seen operating as passenger, cargo and research aircraft in some of the most challenging regions.

Aces High has been specialising in providing aircraft for use in the film and advertising industries for the last 30 years. Its aircraft, ranging from a First World War RFC SE5A to a Russian ground-attack helicopter, are also regular visitors to the many air shows during the season. One of the most popular aircraft, which has featured in many films, is its Dakota, N147DC.

Air Atlantique's Classic Flight

Tim Badham

Air Atlantique has been the owner/operator of Dakotas for many years and the individual histories of its three aircraft can be found in Appendix II. Tim Badham, a volunteer with the Ground Support Team of Air Atlantique's Classic Flight, provides an overview of the company and its aircraft.

In the immediate post-war years many British airlines used Dakotas for short- and medium-haul operations. Of the numerous operators, the last one to utilise a sizeable fleet was

RIGHT The Northamptonshire countryside can be glimpsed below the cloud from the cockpit of G-AMRA, crewed by John Dodd, pilot, and Ben Cox, co-pilot, en route to Coventry in August 2010. *(Tim Badham)*

ZA947's running costs (at 2011 values)

Fuel	approx £1.80 per litre (average fuel load 400gal per trip)
Oil	for a 45gal drum, £720
Tyres	£1,195 each
Propeller	£8,500 for overhaul
Engine	new engine £38,750; exchange engine £34,500; overhaul £28,300, plus any special machining and modifications needed at £37 per hour
Crankshaft	£3,075
Master rod bearing shells	£460 each
Master rod	£460 each
Complete repaint	£60,000–£80,000
Major servicing	£135,000, excluding any rectification work

LEFT Aces High's Dakota, N147DC, on arrival at North Weald. *(Aces High)*

Dakota survivor – N147DC

C-47A Dakota, 42-100884, was built in 1942 at Long Beach. She was delivered to the USAAF in December 1943 and (as Mk III, TS423) transferred to the RAF's No 1 Heavy Glider Maintenance Unit at RAF Netheravon in September 1944. She then passed to 436 (RCAF) Squadron at RAF Down Ampney in August 1945.

Post-war

After participating in the Berlin Airlift, TS423 went to Short Brothers, Scottish Aviation, Marshalls and Ferranti. At Ferranti she acquired a non-standard nose that would have been the envy of Jimmy 'Schnozzle' Durante! Inside this nose, which singled her out from her 11,000 sisters, Ferranti fitted air pass radar for the Lightning fighter and a hydraulically operated gun turret. Underneath she acquired sonar buoy-dropping equipment.

In 1969, with a surprisingly low number of hours on her airframe (around 3,000), TS423 was delivered to the Ministry of Technology at RAE, West Freugh, where, nine years later, named 'Mayfly', she was still stored. The decision was taken by the MOD to scrap her

on the Catterick fire dump. Mike Woodley of Aces High managed to rescue her from this fate with the aid of the Imperial War Museum and Lord Onslow.

On 14 September 1979 TS423 was officially christened G-DAKS and given extensive cosmetic surgery to restore her Douglas nose at Duxford. G-DAKS made her debut with Yorkshire Television as 'Vera Lynn' of Rusking Air Services in the TV series *Airline* in 1979.

Her film and television credits range from *Poirot* to *Tenko*, *Indiana Jones*, *The Dirty Dozen*, *Memphis Belle*, *Saving Private Ryan*, *Band of Brothers*, *The Da Vinci Code* and many others. She has recently returned from her latest movie, *Red Tails*, filmed in Prague in April 2010, and her latest television appearance was in *Land Girls* for BBC1, screened in 2011.

She remains a young lady of 69 years. With just over 3,500 hours total flying time, N147DC is probably the lowest-houred Dakota flying in the world today. Aces High still operates her and she continues to be used in films. You can see her on the air-show circuit with her unrestored, original Second World War cockpit and interior.

Air Atlantique, basing its aircraft initially in Jersey, but moving later to Coventry. Directly, or through associated companies, this organisation, over time, has owned or operated 13 different airframes. The members of the fleet became a common sight, flying the skies in a variety of roles ranging from freight hauling and passenger flying through to radar testing and anti-pollution work.

At one stage, seven of these Dakotas were adapted to carry large tanks inside their fuselages and spray booms were fitted externally to enable them to fulfil a contract to conduct anti-pollution work to combat oil spills at sea. They had to be on standby in readiness, sometimes at remote locations.

The fleet has since been reduced in size and the Dakota is not now used in the UK for commercial passenger operations. Several of the ex-Air Atlantique fleet were released for static preservation at museums and RAF bases around the country; however, three of the fleet were retained and remain active, based at Coventry where Air Atlantique now maintains the Classic Flight of over 20 operational historic aeroplanes.

The three residual Dakotas – G-AMPY, G-AMRA and G-ANAF – still earn their keep through involvement in air-show display work, radar equipment testing and anti-pollution spraying work. All were constructed as C-47B-DKs, manufactured during the Second World War at Douglas's Oklahoma City plant. They were originally destined for the USAAF, but were diverted for use by the Royal Air Force, which called the type the Dakota. Post-war, all three served nascent UK airlines, operating as workhorses with a number of other companies before joining Air Atlantique.

ABOVE G-AMRA in Air Atlantique colours, displaying over Sywell in August 2010. *(Tim Badham)*

BELOW G-AMRA runs up on the ramp with the Classic Flight's Jet Provost T3 in the background at Coventry on the night of 11 December 2010. *(Tim Badham)*

BOTTOM Co-pilot John Goudy peers out of the cockpit as Dakota G-AMPY is taxied onto the Coventry ramp after returning from a demonstration of its anti-pollution spraying capability in March 2009. The aircraft finish may look a little worn, but this is deceptive. The airframe had been over-painted in a temporary colour, using washable paint, but heavy rain and spray from low-level work over the sea had begun to wash some of this off, revealing its permanent scheme underneath. *(Tim Badham)*

Chapter Five

The Crew's view

ZA947 is a key member of the BBMF's fleet and performs a number of roles. The crewmembers are highly trained RAF personnel who volunteer to crew the aircraft in their own time during the air-show season and for special events throughout the year. Squadron Leader Stu Reid was the pilot of ZA947 from 1999 to 2010. In this chapter he shares with us the handling of the aircraft, the different roles of the crewmembers and the display techniques required to show off the aircraft to the public.

OPPOSITE ZA947 in 267 Squadron markings.
(BBMF/Crown Copyright)

Squadron Leader Stu Reid, RAF

I have had the good fortune to serve on the Battle of Britain Memorial Flight (BBMF) as a pilot of DC-3 Dakota, ZA947, and Lancaster BI, PA474, for 11 years. For the majority of that time I was the 'Bomber Leader', which, in addition to routine public appearances with both aircraft, required me to train all the new pilots destined to fly the two multi-engine types with BBMF. My qualifications for the position were being in the right place at the right time and being an unabashed aviation enthusiast who was determined to be one of the privileged few to fly two of the most iconic artefacts of world aviation history. I happened also to have a wide range of RAF flying experience and, at the time of my joining the BBMF, was a Qualified Flying Instructor on the E-3D Sentry AEW Mk 1, an aircraft that while so obviously different from the two BBMF multi-engine types is, at the same time, surprisingly similar in so many other ways.

PLEASE NOTE: It must be stressed that the text is not an authoritative reference to operate the Dakota or similar aircraft. The narrative is a perspective of flying ZA947 through knowledge gleaned from a variety of sources and from personal practical experience.

The aircraft

Dakota ZA947 is unique among contemporary RAF aircraft as being one of only two multi-engine 'tail-dragger' types in RAF service, the other being its BBMF stable mate Lancaster BI, PA474. With two large main wheels at the front and a single, smaller tail wheel at the rear of the fuselage, both aircraft exhibit the unique handling attributes of the 'tail-dragger' undercarriage configuration, but on a much larger scale. As will be discussed in more detail later, it is no accident that the undercarriage on modern military and commercial aircraft is arranged with one or more nose wheel(s) at the front and bigger, heavier main wheels towards the rear in the so-called tricycle configuration, but it is an entirely different matter when the same physics are let loose on a 'tail-dragger' type. Even Isaac Newton would have said, 'You don't want to put the wheels on like that', had he been around at the time of the aviation pioneers, but, like them, he too would have had to make do with the best that technology had to offer at the time. It was not until the end of the Second World War that aviation technology flourished to the extent that, with one or two exceptions, the 'tail-dragger' undercarriage configuration was consigned largely to history.

ZA947 is the training aircraft for all the BBMF multi-engine pilots who will fly the Dakota as a display aircraft in its own right and operate also as a co-pilot on the BBMF Lancaster PA474. Although all Dakota pilots qualify as Lancaster co-pilots, not all Dakota pilots will necessarily move to the left seat of the Lancaster. The Dakota is a docile aircraft, yet it can be very demanding. It is true to say that the Dakota 'learning curve' is extremely steep and relies heavily on the experience, knowledge and adaptability of the trainee pilot in coming to terms with its handling qualities. 'Come to terms' is the most appropriate turn of phrase in the case of ZA947, as it is never possible to be at ease with a 'tail-dragger' and the pilot must remain vigilant at all times. Special consideration must be afforded to operating the aircraft owing to the unique handling characteristic of the type and the pitfalls of any complacency. Unlike most modern types, in certain phases of flight the aircraft is very unforgiving of either a moment's inattention or mishandling if basic procedures are not followed.

It must be remembered also that the Dakota engine is very different from those encountered on most contemporary aircraft types. The engines are, for 1930s technology, generally very reliable, but operating limits and restrictions really are limits and the consequences of non-compliance will usually result in problems or failure. Big radial piston aero-engines utilise the power obtained from many components 'thrashing' around inside, and in the case of the Pratt & Whitney Twin Wasp R90-D, from 14 large, heavy pistons reciprocating at high speed. The breakdown or failure of such components within the engine will inevitably necessitate an engine shutdown to preclude further damage or result in engine failure as the failed component takes its toll on other internal working parts of the engine.

Dakota pilot training

There are four pilot training sorties flown by pilots on conversion to the BBMF Dakota. The first two are the Basic Conversion (Convex) sortie (CV1), which covers all aspects of basic aircraft operation and handling, and the Asymmetric Convex sortie (CV2), which introduces airborne engine shutdown, propeller feathering, asymmetric handling, unfeathering and restart procedures. The trainee pilot is usually sent solo following completion of CV2 on an engines-running handover of command. The first solo sortie, designated CV3, allows the trainee to operate the aircraft as a single pilot, with a navigator occupying the right seat, for approximately 30 minutes, and consists largely of normal visual circuits. Finally, the Display Convex sortie (CV4) introduces the trainee pilot to all aspects of the Dakota display profile and planning considerations. All new pilots undertake the four convex sorties in succession and, on qualification, the pilot will complete routine Pilot Continuation Training (PCT) sorties periodically to maintain aircraft handling skills, display proficiency and refine flying accuracy.

Pre-flight

Meteorology and pre-flight briefing

Before every flight, the forecast weather conditions are checked as they play a big part in the conduct of the sortie. Although the BBMF Dakota operations are always under Visual Flight Rules (VFR), the surface wind has the greatest influence on Dakota handling on the ground and, in particular, during the take-off and landing phases of flight as will be discussed later. ZA947, like most aircraft, is also susceptible to icing, but, unlike most contemporary aircraft types, is not equipped with any functional form of airframe anti-icing or de-icing systems. Consequently, flight in icing conditions must be avoided.

Following the weather brief, the crew will assemble for their pre-flight briefing. The format

Sortie briefing guide

A Sortie overview	Outline plan of events
B Call sign	'Memorial Flight/Lancaster/Hurricane/ Spitfire 1/Spitfire 2, etc.
C Aircraft details	Who, in which aircraft
D POB	Pax, pax escort, seat assignment, turrets
E Sortie timings	Start, taxi, take-off, venue, land
F Weather	Take-off, en route, venue, destination, divs
G Airfield details	Freqs, taxi route, R/W, aids
H Fuel	Load, required, bingo/div, 'combat'
I Performance	Take-off weight versus max take-off weight, airfield obstructions
J Check-in frequency	
K Start-up/taxi/run-up area/take-off order/departure	
L Lose plan	
M Nav brief	Flight plan, route, S.alt, terrain, en route diversions, dip clearance, NOTAMS, Royal Flight, pubs, TAPS, charts, time CX
N Venue	Booked, approved (PPN), handling agent, crash cat
Weather – wind	
	Freqs, times, display order, orientation vs crowd
Line – emergencies	
	Breakout direction, emergency divs, POC
O Arrival	Left/right/opposition break
P Special procedures	e.g., ceremonial, air-to-air photography, etc.
Q Requirements	Practise display, asymmetric
R Latest order	Signatures, amendments
S Safety points	FOD, noise, loose articles, mobile phones
T Authorisation/ signed out	Self-auth, civilians, unusual activity
U Outbrief time	
V Next event	Planning, fuel, report times
W Questions & points	

of most BBMF briefings is the same, but the content will vary to suit the task. The following briefing items are an extract from the BBMF Bomber Aircrew Aid.

Trainee-pilot briefing

Training sorties require a more in-depth analysis of the sortie profile, with additional emphasis on exercise content so that the trainee pilot is in no doubt as to what to expect. The Board Brief for the Dakota Basic Convex sortie (CV1) is shown below.

The idea of the Board Brief is to enable the trainee pilot to return later and review the sortie profile and content. The brief is set out in the training *objectives*, which should be self-explanatory and introduce the new pilot to special *considerations* that are unique to the Dakota. They are summarised as follows:

Technology (1930s) – Unlike modern aircraft, there is little automation and there are few fail-safe systems – the *pilot* is the safety break.

Tail-dragger – The aircraft will exhibit unique handling characteristics. These are briefed to the pilot separately and are touched on later in this chapter.

Propellers – Propellers are highly dangerous, especially when attached to a piston engine. Always consider the prop plane as 'live' – walk round them NEVER through them.

Radial engine – The engines require careful handling at all times – look after them and they will look after you.

Noise – The aircraft is virtually silent until start then the engine noise can be unexpectedly intrusive.

The *airmanship* topics relate to both safety limitations and the correct operation of the aircraft in flight:

Lookout – The greatest threat to an aircraft in flight is another aircraft on a conflicting flight path with a risk of collision. The pilot must maintain a good lookout for other aircraft at all times. It is not easy in the Dakota.

Limits – There are numerous limitations that apply to the BBMF Dakota and they *must* be adhered to otherwise the long-term integrity of the aircraft will be compromised and premature failure of components will occur.

Engine handling – The engines need to be 'nurtured' in flight and controls must be manipulated carefully as explained later in the chapter.

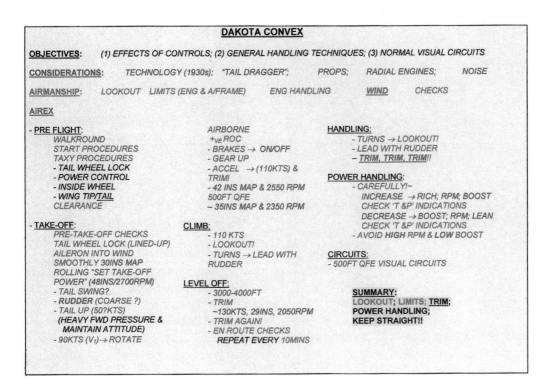

DAKOTA CONVEX

OBJECTIVES: *(1) EFFECTS OF CONTROLS; (2) GENERAL HANDLING TECHNIQUES; (3) NORMAL VISUAL CIRCUITS*

CONSIDERATIONS: *TECHNOLOGY (1930s); "TAIL DRAGGER"; PROPS; RADIAL ENGINES; NOISE*

AIRMANSHIP: *LOOKOUT LIMITS (ENG & A/FRAME) ENG HANDLING* **WIND** *CHECKS*

AIREX

- **PRE FLIGHT:**
  WALKROUND
  START PROCEDURES
  TAXY PROCEDURES
  - **TAIL WHEEL LOCK**
  - **POWER CONTROL**
  - **INSIDE WHEEL**
  - **WING TIP/TAIL**
  CLEARANCE

- **TAKE-OFF:**
  PRE-TAKE-OFF CHECKS
  TAIL WHEEL LOCK (LINED-UP)
  AILERON INTO WIND
  SMOOTHLY 30INS MAP
  ROLLING "SET TAKE-OFF
  POWER" (48INS/2700RPM)
  - *TAIL SWING?*
  - *RUDDER (COARSE ?)*
  - *TAIL UP (50?KTS)*
    (HEAVY FWD PRESSURE &
    MAINTAIN ATTITUDE)
  - *90KTS (V_1)→ ROTATE*

AIRBORNE
+$_{VE}$ ROC
- *BRAKES → ON/OFF*
- *GEAR UP*
- *ACCEL →(110KTS) &*
TRIM!
- *42 INS MAP & 2550 RPM*
500FT QFE
- *35INS MAP & 2350 RPM*

CLIMB:
- *110 KTS*
- *LOOKOUT!*
- *TURNS → LEAD WITH*
RUDDER

LEVEL OFF:
- *3000-4000FT*
- *TRIM*
~130KTS, 29INS, 2050RPM
- *TRIM AGAIN!*
- *EN ROUTE CHECKS*
  REPEAT EVERY 10MINS

HANDLING:
- *TURNS → LOOKOUT!*
- *LEAD WITH RUDDER*
- *TRIM, TRIM, TRIM!!*

POWER HANDLING:
- *CAREFULLY!~*
  INCREASE → RICH; RPM; BOOST
  CHECK 'T &P' INDICATIONS
  DECREASE → BOOST; RPM; LEAN
  CHECK 'T &P' INDICATIONS
- *AVOID HIGH RPM & LOW BOOST*

CIRCUITS:
- *500FT QFE VISUAL CIRCUITS*

SUMMARY:
LOOKOUT; LIMITS; TRIM;
POWER HANDLING;
KEEP STRAIGHT!!

Wind – The pilot must take account of the surface wind when operating the aircraft either on or near the surface.

Checks – All checks must be completed carefully and methodically. They take time – DO NOT rush them.

The *Airex* sets out the conduct of the sortie, so that the training objectives are achieved in a logical, step-by-step and progressive sequence. The brief is concluded with the punch lines *summary*, which highlights the main learning points of the exercise. The Asymmetric and Display Convex Sorties also utilise a similar Board Brief to set out the various training objectives and main learning points.

Pre-flight administration

Following the pre-flight briefing, the sortie details are entered in the Flight Authorisation Sheet (RAF Form 1575B), a legal document that sets out exactly what is to be done during the flight, by whom, when and in which aircraft. The final out-brief is completed and the aircraft commander checks the aircraft servicing

BELOW **Waiting to load and start the aircraft ready for another display.** *(BBMF/Crown Copyright)*

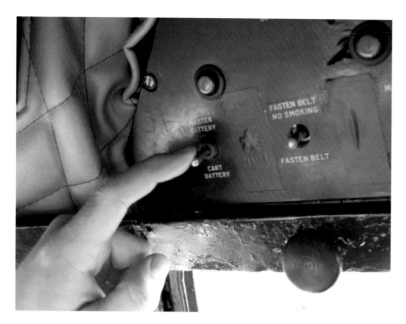

and maintenance document (F700) while the rest of the crew make their way to the aircraft to commence the pre-flight checks. The F700 is a summary record of technical information, including leading particulars, routine maintenance, completed rectification work, outstanding rectification work, record of flying hours and any deficiencies, which will not affect safety of flight, but may impact on the conduct of the sortie. Having

scrutinised the F700 carefully and signed as having 'accepted' the aircraft for flight, the aircraft commander proceeds to the aircraft to join the rest of the crew.

Aircraft exterior inspection

On arrival at the aircraft the pilot confirms that there is sufficient residual hydraulic pressure (500psi) to withdraw the two main undercarriage pins and requests the pitot heaters to be selected ON. The walk-round commences from the main access door on the left-rear side of the fuselage, usually in a clockwise direction. The aim of the exterior check is to confirm the final integrity of the aircraft before flight.

A general visual inspection of the aircraft exterior is carried out, including security of the fuel tank filler caps, the condition of the aircraft, skin, flight controls, engine, wheels and tyres and that the pitot-probe heaters are serviceable, to mention but a few items. Most pilots use the walk-round to prepare privately for their forthcoming flight and to familiarise themselves with the aircraft in their charge. There is something special about touching the aircraft surface with your hands before flight, it is an opportunity to become acquainted with one another in much the same way as a rider takes

charge of a horse before running out. A pilot communicates with an aircraft and an aircraft communicates with a pilot in many ways and I'll bet there isn't a pilot out there who doesn't talk to an aircraft, one way or another, during a pre-flight walk-round and while in flight.

Starting the engines

The engine start procedure is set out in the STARTING CHECKS and is followed carefully to ensure an attempt to start the engines is effected safely and using the correct sequence of engine control selections to give the greatest chance of success. The number two (right) engine is started first; there are a few reasons for this, the main one being that piston engines require some minutes to 'warm up' after start, and because each engine must be started sequentially. In the early days of passenger flight, passengers would board the aircraft through the main fuselage door, usually on the left side. As the passengers were boarding, the right engine was started to both save time and not subject the boarding passenger to 'prop-wash' and engine exhaust. The number one (left) engine was started as soon as the passenger door was closed, thereby minimising the delay before taxi and the risk of fumes being blown into the cabin.

Big radial piston aero-engines can be quite temperamental to start. Some engines are started using an electric starter motor that drives the engine directly. However, a great deal of electrical energy is required to enable the starter motor to turn the engine over. An alternative starting system, which requires less electrical energy, employs a starter motor/flywheel mechanism. ZA947 is equipped with two Pratt & Whitney R90-D engines, which utilise an inertia starter-type mechanism incorporating a starter motor and flywheel. To enable the motor flywheel to acquire sufficient rotational energy to turn the engine, the starter motor is energised for 15 seconds using a three-position STARTER ENERGISE switch. The switch is spring-loaded to the centre OFF position and must be held in either the RIGHT (up) position to energise the right starter motor or the LEFT (up) position to energise the left. After 15 seconds have elapsed, and while maintaining the ENERGISE switch in the

ABOVE Operating the ENERGISE and MESH switches. *(BBMF/ Crown Copyright)*

appropriate position, a three-position STARTER MESH switch is operated to engage the starter motor/flywheel with the respective engine to turn it over. As with the ENERGISE switch, the MESH switch is spring-loaded to the centre OFF position also and is held in either the RIGHT (down) position to mesh the right starter motor with the right engine or the LEFT (up) position for the left. Both switches are operated from the right seat with the various selections, including switch release either after start or to stop the start, being made on the pilot's command.

Regardless of the type of starter system used, the starter motor, or start motor/flywheel, is unable to turn the engine at sufficient rpm to utilise the normal carburettor fuel supply for start-up as is used when the engine is running. The motor/flywheel is very heavily loaded when it is turning the engine and the slow rotation draws minimal air into the cylinders. If the carburettor mixture control is opened with the engine rotating on the starter motor/flywheel, an over-rich mixture is admitted to all cylinders, which will not ignite easily and greatly increases the risk of engine damage through so-called 'hydraulic-ing'. The damage occurs if oil or fuel pools in one or more of the lower-most cylinders, with insufficient clearance between the piston crown and cylinder head, forcing the engine to either stop suddenly, reverse direction or fracture the cylinder head.

Consequently, to assist with engine starting

a priming system is used to administer a small quantity of fuel to a few of the upper-most cylinders which, when ignited, increases engine rotation speed to an rpm sufficient for carburettor fuel to be admitted to all cylinders and enable a self-sustaining idle rpm to be achieved. In the case of the Pratt & Whitney Twin Wasp, priming fuel is fed to the upper four cylinders of the rear bank of seven under electric booster pump pressure through a valve controlled by a PRIME switch on the flight deck. There is one, three-position PRIME-OFF-OIL DILUTION switch for each engine, which is spring-loaded to the centre (OFF) position. Both switches are operated as necessary by

the pilot and are held in the down (PRIME) position against spring pressure to activate the primer valve. Operation of the valve is confirmed by a slight reduction in gauge-indicated fuel pump fuel pressure to the respective engine. The switch may be held, also, in the up (OIL DILUTION) position; however, the oil dilution facility is disabled on ZA947, but not necessarily on civilian-operated aircraft. The difficulty for the pilot is judging the use of the primer system in terms of both when to prime and for how long, positioning the throttle correctly for start and timely selection of both the ignition switch and carburettor fuel control – apart from that it's pretty straightforward!

With the initial starting checks complete, the right booster pump running, and having obtained start clearance from Air Traffic Control (ATC), the stopwatch is set to zero and the right seat occupant prepares to operate the right engine START switch on the command of the pilot. At this stage, the pilot sets the IGNITION MASTER switch on and operates the right engine primer switch three or four times, each for about one second, to allow fuel under booster pump pressure to flow to the four upper cylinders. It is very easy to over-prime and flood the upper cylinders, the result of which is the same as for under-priming, in that the engine will fail to start. If in doubt, practical experience has shown the best option is to under-prime the engine prior to start and then operate the primer again as the engine starts to turn under influence of the starter motor/flywheel.

During start-up, the majority of engine controls are operated from the left seat and require considerable dexterity for the start to be successful – a bit like learning a 'hill start' in a car for the first time. Having primed the right engine, the pilot covers the right engine ignition switch with his right hand and is poised to start the clock with his left hand. The pilot calls 'Energise' and, at the same time, starts the clock then repositions his left hand to cover the engine primer switch. Fifteen seconds after the right seat occupant operated the right engine START switch, the pilot calls 'Mesh' at which point, and while maintaining pressure on the START switch, the right seat occupant operates the right engine MESH switch to engage the motor/flywheel with the engine and

then observes the right engine in anticipation of it turning over. Depending on whether or not the engine was hand-turned during the pre-flight servicing, the right seat occupant waits for the passage of the appropriate number of 'blades', either three for a turned engine or nine otherwise, before calling 'Three/nine blades'. On the call, the pilot selects the right engine ignition switch to BOTH and repositions his right hand on the right engine MIXTURE/IDLE CUT-OFF control lever. If the engine catches as the ignition switch is selected to BOTH, the pilot 'blips' the PRIME switch momentarily, once or twice, with his left hand, to keep the engine firing over and simultaneously moves the MIXTURE/IDLE CUT-OFF control lever to the AUTO-RICH position with the other hand, 'blipping' the PRIME switch again as necessary until the right engine starts to run smoothly on fuel supplied from the carburettor. After a few seconds, as the engine rpm starts to increase and stabilise, the pilot transfers his right hand from the mixture control lever to the right engine throttle and calls 'On the throttle' at which point the right seat occupant releases both the right engine START and MESH switches. The pilot advances the throttle carefully to bring the right engine to a fast-idle of 900–1,000rpm to prevent heat damage to the spark plug elbow insulation and reduce the risk of spark plug oil fouling. The start procedure is then repeated for the left engine.

Failure to start

There are many reasons why an engine will fail to start but, in the absence of a technical malfunction, failure to start first time is usually due to a priming irregularity. After the starter motor/flywheel-operating limit has elapsed, both the START and MESH switches are released and the MIXTURE/IDLE CUT-OFF control lever is confirmed in the IDLE/CUT-OFF position. When the engine stops rotating, the engine ignition switch is selected OFF, the booster pump is turned off. After the starter motor cooling time has elapsed, another start may be attempted.

After start

With both engines running at 900–1,000rpm, the AFTER START checks are completed. A checklist item particularly noted is the static bleed reading, which is the static atmospheric

LEFT Operating the right hand engine magneto switch. *(BBMF/Crown Copyright)*

BELOW Moving the mixture lever to the AUTO RICH position. *(BBMF/ Crown Copyright)*

LEFT Moving the mixture control lever to the OFF position if the engine fails to start. *(BBMF/ Crown Copyright)*

pressure reading indicated on the engine manifold pressure gauge when the static bleed position is selected on the gauge input selector. The figure is usually about 30in and is the throttle setting used during the engine power checks, which are completed before take-off. The engines must be monitored carefully as they warm up and the throttle adjusted regularly to maintain 1,000rpm as the engine oil and cylinder head (CHT) temperatures increase. After a cold start, the viscosity of the oil reduces as the engine warms and the rpm will rise steadily without timely adjustments to the throttle position, possibly exceeding the cold engine 1,200rpm limit. The engine requires a minimum CHT of 120°C and oil temperature of 40°C before the engine may exceed 1,200rpm. As soon as one of the engines has achieved the minimum temperature parameters, the rpm is increased to 1,300rpm to bring a generator on-line to charge the aircraft batteries.

Taxi procedure

Being a 'tail-dragger', the Dakota is not an easy aircraft to taxi, particularly if the surface wind exceeds 10kts, because the tail-wheel configuration makes the aircraft prone to weather-cocking. The subject of tail-dragger handling characteristics is beyond the scope of this book, but broadly speaking, under certain circumstances, directional instability renders the aircraft difficult, if not impossible, to control. The undercarriage configuration of the aircraft

is such that, during phases of operation on the ground, any directional deviation from the axis of motion may result in loss of control and the aircraft ground looping under the laws of natural physics. It is always important to appreciate the effect of the surface wind on maintaining directional control and to anticipate the use of both rudder and differential brake, combined with differential engine power, to steer the aircraft on the ground. Because the rudder is used for directional control on the ground and needs to be effective, even at low indicated airspeeds, it is disproportionately large in relation to the rest of the aircraft vertical stabiliser and compared with modern aircraft designs configured with a conventional, by modern standards, nose-wheel. The size of the rudder is such that, in some surface-wind scenarios, the foot-loads are extremely high and it can be difficult to prevent the rudder from being blown to full-scale deflection.

Tail-wheel lock

In the Dakota directional control is helped, albeit to a limited degree, through use of a lockable tail wheel. The tail wheel can be locked in position when it is aligned directly with the aircraft fuselage axis. The lock is operated by a lever under the central pedestal on the flight deck, which allows a pin, acting under spring pressure, to be engaged in a receptacle on the tail-wheel strut, but only when it is aligned with the aircraft fuselage. The locked tail wheel assists with maintaining directional control as it is unable to caster under the influence of the wind or inertial factors acting on the aircraft, which would otherwise allow the aircraft to swing against the desired direction of movement. However, it is only beneficial when taxiing in a straight direction and must be disengaged when necessary to turn the aircraft on the ground. Moreover, it can be very difficult, if not impossible, to disengage the tail-wheel lock if it is subject to anti-lateral load as is applied when surface wind is acting against the rudder. We have all experienced one of those 'Oooh-lummy', or words to that effect, moments when the tail-wheel lock refuses to disengage as the edge of the taxiway beckons.

With the taxi checks complete and the temperatures of both engines above the

BELOW Disengaging the tail wheel lock prior to taxiing. *(BBMF/ Crown Copyright)*

minimum, the aircraft is ready to taxi. Taxi clearance is obtained from ATC, the tail-wheel lock is disengaged and, with the wheel brakes applied, the parking brake is released. The aircraft is easier to control at slow speed with the engine at 1,000–1,300rpm and reducing inside engine rpm combined with inside main-wheel brake when required to turn. As the aircraft moves forward, the wheel brakes are each checked for correct operation momentarily. Anticipating the effect of wind, leading with rudder in a turn, differential brake combined with differential engine rpm is used to steer the aircraft on the ground.

Engine power checks

In the same way both the accelerator and gearbox on a road vehicle optimise engine power with driving conditions, the engine controls on ZA947 are designed to optimise engine power in relation to flight conditions. The throttle control, as with an accelerator, increases fuel flow to the engine to increase engine speed and power output. The rpm lever, as with a gearbox, varies the propeller pitch to change the air load on the propeller and, thus, engine rpm for a given throttle setting enabling high engine power output to be achieved combined with low rpm in normal flight regimes.

Having taxied to a clear area, well away from any obstructions and possible sources of debris, the aircraft is positioned into wind in preparation for the pre-flight engine RUN-UP CHECKS. The purpose of the checks is to ensure that the engines are operating correctly at higher power settings, as used in flight, but before flight so that any irregularity can be investigated safely. It is unwise to fly without having first ensured that both the engines and engine controls are fully functional and that engine power output is optimal.

With the parking brake applied, the control column is restrained against the back-stop to ensure prop-wash over the tail plane and elevator acts to keep the tail wheel in firm contact with the ground. Also the throttle friction control, a rotary knob under the central pedestal, is applied firmly to make sure that engine vibration does not cause the throttles to retard slowly as the checks are carried out while they are unrestrained by the pilot's hand, thus resulting in erroneous engine indications.

With the rpm control levers in the MAX position, both engines are opened up to 1,700rpm using the throttles. Once stabilised at 1,700rpm, both rpm levers are then exercised rearward simultaneously from MAX to MIN in one smooth, progressive movement. Movement of the levers allows oil under pressure to act against the propeller blade pitch control mechanism in the propeller Constant Speed Unit (CSU), to force the propeller blade pitch to increase. With the throttle position unchanged from that selected to achieve 1,700rpm, the increasing pitch angle of the propeller blades increases the air load on the propeller and the rpm decreases to 1,200–1,300rpm typically. The rpm is permitted to stabilise momentarily before repositioning the rpm levers to the fully forward MAX, again in one smooth, progressive movement, which allows oil under pressure to act on the opposing side of the CSU propeller pitch control mechanism, forcing the propeller pitch angle to reduce. With a corresponding reduction in the air load, the engine rpm increases readily and returns to approximately 1,700rpm.

The procedure is then repeated for a second time, the purpose of which is to circulate warm engine oil through each propeller CSU, purging the CSU components before throttle setting and rpm are compared at the reference parameters. Such is the effect of exercising the rpm levers,

BELOW Moving the throttles to do run-up checks. *(BBMF/ Crown Copyright)*

it is usually necessary for the throttle(s) to be adjusted slightly to ensure an accurate 1,700rpm is maintained.

After the rpm levers have been exercised twice, each propeller feathering system is checked in turn for correct operation. At a steady 1,700rpm, the No. 1 (left) engine feathering button is depressed and the ammeters monitored for an increase in current flow to the propeller feathering pump electric motor. The motor forces very high pressure oil into the propeller pitch control mechanism, overcoming CSU oil pressure to increase the propeller pitch and start to feather the propeller. As soon as the increased current flow is observed, the rpm gauge is monitored for a 200rpm reduction as the propeller starts to feather under the influence of high-pressure oil from the feathering pump. In the same way as with exercising the rpm levers, the engine rpm reduction is in response to the greater air load applied to the propeller blades as the pitch is increased. As soon as a drop of 200rpm is observed, the feathering button is pulled out, stopping the feathering pump and ceasing the flow of high-pressure oil to allow CSU oil pressure to control propeller blade pitch normally. The rpm then starts to rise in response to reducing propeller air load as oil pressure within the CSU acts to reduce propeller blade pitch and the engine recovers to 1,700rpm.

The right engine propeller feathering system is then checked for correct operation in the same way, following which both engine rpm levers are exercised simultaneously from MAX to MIN and MIN to MAX for a third time. With both engines running, once again at 1,700rpm, the right engine is throttled back to 1,100rpm and the left engine throttle advanced further until the left engine manifold pressure (MAP) indicator reads 'static' MAP as was noted in the AFTER START CHECKS. The rpm at the static MAP throttle setting, usually approximately 2,400rpm, is noted then each magneto is checked by operating the respective magneto selector switch from BOTH to No. 1, noting the rpm drop, then back to BOTH, then from BOTH to No. 2, noting the rpm drop, then back to BOTH, ensuring the mag drop at each individual magneto position does not exceed 65rpm. A mag drop of greater than 65rpm is indicative of an ignition problem, such as a defective spark plug, which will affect engine power output and must be investigated before flight. With the check at static MAP complete, the engine is throttled back smoothly to 1,100rpm and the static MAP power check completed for the other engine, after which both engine throttles are closed smoothly and the idle rpm is checked at 600–700rpm.

With the engine RUN-UP CHECKS complete, the aircraft is taxied to the runway holding point for the TAKE-OFF CHECKS, which are a final series of checks to ensure that the aircraft and all systems are configured appropriately for flight. Some systems are either potentially hazardous to bystanders or require considerable electric power and are, therefore, activated only immediately prior to take-off.

In-flight

Take-off

With the TAKE-OFF CHECKS complete and take-off clearance obtained from ATC, the aircraft is taxied onto the runway, paying particular attention to the wind direction and strength, and the RUNWAY CHECKS completed. It is worth pointing out, at this stage, that the Dakota, in common with other tail-dragger aircraft types equipped with piston aero-engines, is inclined to swing as engine power is increased. The aircraft swings under the influence of both the prop-wash over the rear of the fuselage and the gyroscopic effects of the propeller blades rotating at high rpm as the tail is raised during the take-off roll. The direction of swing is a function of propeller direction of rotation, which, in the case of the Dakota, causes the nose to swing to the left (tail to the right) as the pilot senses it, when the engine power is increased for take-off. A degree of anticipation, combined with prompt, coarse, directional flight control inputs and smooth, progressive throttle handling, inevitably results in an uneventful, though sometimes interesting, take-off roll.

However, there is another variable, which can make the take-off (and landing) far more demanding for the pilot: surface wind. It is very important to anticipate the wind's likely

influence when the 'number' is passed by ATC in the take-off clearance. A headwind component for either take-off or landing is desirable as it assists with both aircraft directional control and runway performance. Consequently, a tailwind component is unacceptable for ZA947 flight operations. A more difficult scenario to contend with is a surface wind, which results in neither a head- nor tailwind component, but gives a maximum crosswind component especially from the left. As with taxiing, the aircraft will always tend to weather-cock into wind and any crosswind component will act against the exposed aircraft surfaces and, in particular, the large vertical stabiliser or fin and rudder structure. The Dakota crosswind limit is 15kts, which, when combined with the unpredictability of gusty conditions, is the most undesirable and challenging of surface-wind conditions. A crosswind component of 8 gusting 16kts is,

arguably, a more difficult scenario than a steady 16kts across, because of the unpredictable nature of gusty conditions.

The problem with a crosswind from the left is that the combined effects of the wind, prop-wash and gyroscopic effects result in significantly reduced directional control authority at low IAS (Indicated Air Speed), because the fin/rudder must act against all three opposing factors. However, a crosswind from the right mitigates the prop-wash and gyroscopic effects, resulting in the rudder controls having greater authority in retaining directional control of the aircraft.

Against a background of widely differing surface-wind conditions, the pilot will position the aircraft on the runway so as to take advantage of the runway dimensions to compensate for the surface wind. For example, with a prevailing crosswind component from the left, the pilot will position the aircraft left of the runway centreline, usually with the centreline outboard of the right

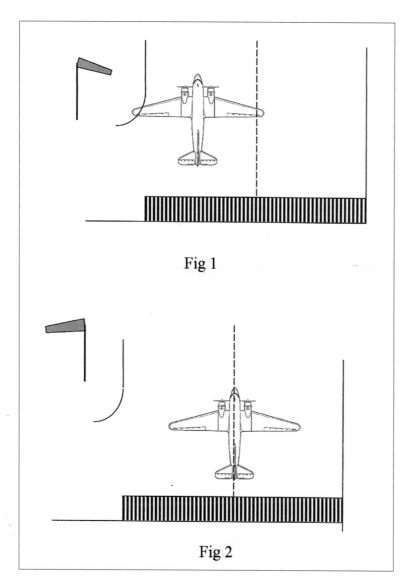

Fig 1

Fig 2

with the runway surface to be effective and it is reduced markedly by a wet runway surface or the aerodynamic lift effects of the airflow over the tailplane, reducing the proportion of aircraft weight borne by the tail wheel.

Therefore, the pilot maintains the control column fully aft with his left hand, so that the prop-wash helps to maintain a downward force on the tail wheel. The wheel brakes are released (if applied when given an ATC instruction to 'Line up and wait') and the pilot opens both throttles progressively to approximately 30in MAP, while maintaining directional control with coarse use of the rudder. If the take-off roll is commenced from the runway position, as in Figure 1, the aircraft will, naturally, drift towards the runway centreline and, at the same time, start to align with the runway centreline under the influence of the increasing prop-wash from the engines acting against the vertical tail and rudder as the throttles are opened.

With the engines stabilised as 30in MAP and all engine indications confirmed in the normal range, the pilot relinquishes control of the throttles to the non-handling crewmember and calls to, 'Select 48in and 2,700rpm.' The pilot places his right hand on the other side of the control column while maintaining the back pressure to keep the tail in contact with the runway as the throttles are opened to take-off power. The pilot has to work hard at this stage to co-ordinate rudder inputs to maintain the aircraft alignment with the runway centreline, while applying aileron into wind to keep the up-wind wing from rising and maintaining the back pressure on the control column until the Airspeed Indicator (ASI) reacts to the increasing airspeed and displays the Indicated Airspeed (IAS) as soon as the ASI starts to read. While maintaining directional control with the rudder combined with the aileron into wind, the pilot pushes forward smartly on the control column to lift the tail off the runway and achieve a level aircraft attitude, with the main wheels maintaining positive contact with the runway.

As discussed previously, the aircraft is inclined to swing to the left under the influence of the propeller prop-wash, an effect that is exacerbated by the propeller gyroscopic effects as the tail is raised on take-off. The huge rudder rapidly gains effect and it is easy for the trainee

engine, and aligned to be pointing one-half to one full width of the runway to the right of the departure end as shown in Figure 1. With a crosswind component from the right, the pilot will position the aircraft on the runway centreline and more or less aligned directly with the aircraft runway as shown in Figure 2.

Regardless of wind direction, the techniques employed on take-off are the same, but the pilot may use the surface wind to advantage for ease of predictable aircraft control. As the aircraft completes the turn onto the runway for take-off, the tail-wheel lock is engaged and the pitot heaters (and landing lights if required) selected ON. As with taxiing the tail-wheel lock assists with directional control during the take-off roll by reducing the tail swing effects. However, it relies on the friction of the tail-wheel tyre in contact pilot to 'over control' the rudder inputs and exacerbate the swing effects. Indeed, rudder control is the most difficult aspect of the take-off for the trainee pilot to grasp. It is a so-called 'V2' relationship, which, in effect, means that rudder authority increases at a greater rate in proportion to the increasing IAS. With the aircraft accelerating, now in a level attitude, the pilot will find that both the rudder inputs and degree of rudder deflection reduce as the increasing IAS also increases rudder authority.

Depending on aircraft All-Up Weight (AUW), the forward force on the control column can be quite high and the forward pressure is maintained to keep the main wheels in contact with the runway as the aircraft accelerates to the Rotation Speed (Vr) of 90kias (Knots Indicated Air Speed). To those unfamiliar with multi-engine aircraft operations, it may seem unusual to keep an aircraft on the runway beyond the minimum IAS required for flight, but the reason is to cater for an engine failure during take-off. In the event of an engine failure during the take-off roll there are two options: either abandon the take-off or continue the take-off on the one or more remaining engine(s). The options are a function of aircraft performance in relation to ambient conditions, runway length and the proximity of obstructions, discussion of which falls outside the scope of this manual. In broad terms, it must be appreciated that depending on the IAS at the time of engine failure, either option is potentially hazardous so, in the case of ZA947, a fixed IAS is used to *assist* the pilot in deciding which option to take. The emphasis is on *assist* the pilot with the decision as safety of flight may dictate that the pilot modifies his course of action.

The reason 90kias is used as the Decision Speed for ZA947 is that although the aircraft will fly successfully on two engines below that IAS, it will not, crucially, fly on one. Below 90kias, there is insufficient control authority for the pilot to counter the asymmetric effects of the failed engine with the other at the take-off power setting required to achieve a safe climb away from the runway and loss of control is inevitable. If the pilot reduces the 'live' engine power to retain directional control, the aircraft will not climb, but it should be possible to land, albeit possibly as a crash-landing beyond the end of

the runway. Consequently, one way or the other, the aircraft will make contact with the surface again and it is just a matter of whether it is under control (survivable) or out of control (not survivable) as it does so.

The non-handling crewmember calls the IAS at 10kias increments from 60kias and at 90kias calls, 'Rotate', at which point the pilot reduces the forward pressure on the control column and applies back pressure to rotate the aircraft off the runway. Once the aircraft is safely clear of the runway the pilot momentarily applies the wheel brakes, being careful not to 'snatch' them, calls, 'My power' and takes control of the throttles from the non-handling crewmember who responds with 'Your power'. The pilot then calls for the gear to be raised.

After take-off

It takes quite a few seconds for the gear to travel to the fully-up position, at which point the pilot reduces power to 42in MAP and 2,550rpm for the initial climb and calls for the AFTER TAKE-OFF CHECKS. The checks ensure the aircraft is configured correctly for the climb and that all systems and indications are functioning normally.

Climb and level-off

Approaching 500ft QFE (atmospheric pressure at aerodrome elevation) with the checks complete, if remaining in the Visual Circuit, the power is reduced to 27in MAP and 2,350rpm, but if continuing the climb out of the Visual Circuit, power is reduced 35in MAP and 2,350rpm. Unlike Lancaster PA474, the engines are not equipped with Automatic Boost Control (ABC), which would maintain the MAP setting at that selected by the pilot as the aircraft altitude increases. Consequently, the pilot must continually adjust (increase) the throttle setting during the climb to compensate for reducing air pressure with increasing altitude. At the desired level-off altitude, the pilot selects cruise power of 29.5in MAP, 2,050rpm and moves the MIXTURE/IDLE CUT-OFF control lever to AUTO-LEAN position and calls for the EN ROUTE CHECKS, which configure the aircraft for prolonged cruise flight. If completing a General Handling exercise, or one of the Convex sorties, the power is manipulated in accordance with the requirements of the sortie profile.

Power settings

Take-off (maximum) power	48in MAP/2,700rpm/AUTO RICH
Maximum except take-off (METO) power	42in MAP/2,550rpm/AUTO RICH
Maximum continuous power	35in MAP/2,350rpm/AUTO RICH
Cruise power	29½in MAP/2,050rpm/AUTO LEAN

Engine power control

One of the main hurdles the trainee pilot must come to terms with in flying ZA947, is engine power control. The power setting must be selected both carefully and accurately and the engine controls must be operated in the correct sequence when either an increase or decrease in engine power is desired. Strict adherence to mixture, throttle (MAP) and rpm setting must be observed. Failure to make the correct selection in the correct order will result in significantly reduced engine power output combined with a greatly enhanced risk of either engine damage or failure.

BELOW ZA947 in her 233 Squadron markings. *(Jarrod Cotter)*

Fortunately, there are only four principal power settings used in flight, each of which optimises engine power output in relation to throttle and rpm settings. The power settings are listed at left.

As discussed previously, the three primary engine power controls for each engine is the THROTTLE lever, the rpm lever and MIXTURE/IDLE CUT-OFF lever and the desire to either increase or decrease in power will dictate the sequence of lever movement. Although simple in concept, it is very easy to manipulate the wrong lever in the wrong sequence and it is important for the trainee pilot to become accustomed with the engine controls at an early stage. Initially, it takes the trainee a good minute or so to complete the power adjustments but, with practice, accurate power settings are achieved within a few seconds. During the first Dakota Convex sortie (CV1), ZA947 is levelled at approximately 2,000ft after departure and the procedure for reducing engine power to maintain level flight is demonstrated to the trainee pilot. Following the climb after take-off with 35in MAP 2,350rpm and Mixture AUTO-RICH, the instructor demonstrates the correct sequence of engine power lever manipulation to

bring the engines into the cruise configuration. In simple terms, to reduce power the pilot must move the levers from left to right as they are grouped across the throttle control pedestal. Accurate setting of cruise power is important, as it is likely to be maintained for a prolonged period.

The first adjustment is to the throttles, which are retarded to achieve approximately 29.5in MAP. Accuracy is not necessary at this stage, because the next levers to be manipulated are the rpm levers, which are set at 2,050rpm and will, in turn, modify the MAP indication slightly, necessitating a further throttle adjustment for refinement. The easiest way to achieve an accurate 2,050rpm setting is to move both rpm levers to the approximate setting. Then, grasping *either* rpm lever at the mid-point between finger and thumb, while resting the same hand on the throttle control pedestal behind the engine control levers, adjust the position of the rpm lever until an exact 2,050rpm is indicated on the gauge. The other rpm lever is then grasped in the same way and, *without scrutinising the rpm indication for accuracy,* the lever position

is refined or 'tweaked' slightly until there is no discernible audible reverberation between the two engines. The problem is that there is a marginal inaccuracy between each engine rpm indicator needle. If the rpm indications are matched exactly, the crew will be faced with prolonged reverberation of the engines in the cruise as the respective rpms move in and out of phase with each other. With the engine rpms synchronised audibly, the mixture control levers are selected to the AUTO-LEAN position for optimal fuel consumption and the EN ROUTE CHECKS are completed.

To increase power, the engine levers are adjusted as grouped from right to left across the throttle pedestal. First, the mixture levers are moved to the AUTO-RICH positions, then the rpm levers advanced to either 2,350, 2,550 or 2,700rpm as desired and finally the throttle levers opened to the maximum setting corresponding to the preselected rpm of either 35, 42 or 48in MAP respectively. Over-boosting the engine, that is too high a MAP setting for the preselected rpm, even if only for a few seconds, will cause significant engine damage.

ABOVE The view in the cockpit as ZA947 climbs away from Coningsby. *(BBMF/ Crown Copyright)*

Prolonged descent

A protracted descent, requiring a corresponding prolonged reduction in engine power, required careful consideration and planning. Being air cooled, with the throttles at idle in flight, the cylinder head temperatures (CHT) cool-off remarkably quickly. Fortunately, most BBMF flight operations do not require ZA947 to fly at altitudes in excess of 2,000ft, but some activity, including parachute drops and the Flight Test profile, require the aircraft to climb to altitudes that necessitate a prolonged subsequent descent. If a prolonged descent is followed by a sudden increase in engine power, the cylinder heads will be subject to thermal shock, resulting in undue stress and the possibility of damage leading to failure. Consequently, unless safety of flight dictates otherwise, idle engine power is not selected in flight and descents are completed with the engine at reduced, 15–20in MAP, as opposed to idle. The cylinder-head temperature (CHT) must be monitored carefully to ensure it does not fall below the in-flight minimum and the engines must be 'cleared' periodically to purge them of any oil accumulation that may lead to plug fouling.

Another major factor to be borne in mind is the relationship between engine MAP and rpm. Radial engines produce the power for flight through the combustion of fuel in the cylinders forcing the pistons inward to drive the crankshaft and the propeller. Although the engine is designed to withstand the forces associated with fuel combustion and driving the propeller with ease, significant damage to the engine will occur if a 'windmilling' propeller is turning the engine at high speed, as would be the case with the throttle set at idle in flight. The problem arises from the pistons being thrown outward on their respective connecting rods and being subject to high centripetal forces through the engine being turned at high rpm by the wind-milling propeller. The connecting rod components are not designed to withstand the extreme forces acting on them in a direction opposite to that which they are supposed to be subjected.

However, it is possible to use the rpm levers to slow the rotation speed of a windmilling propeller and thereby reduce the forces acting within the engine. As long as the engine is turning, the engine oil pump to the propeller CSU facilitating rpm control is supplying oil. Moreover, by complying with a simple 'rule of thumb' relationship in that the MAP must never be less that the first two figures of the rpm, the rpm levers are used to control engine rpm in relation to the MAP indication and the risk of damage to the engine is minimised. For example, with the engine at 2,050rpm, the throttle must not be less than 20.5in MAP to ensure that the engine remains in a safe operating regime. It follows that to descend with a throttle setting of 15in MAP, the rpm must be reduced to a corresponding figure of no more than 1,500rpm. It must be remembered that this relationship is used only to control engine rpm and maintain CHT in the descent on a serviceable engine. For a failed engine the propeller is feathered and the engine does not rotate.

The final matter of which the trainee pilot must be aware is the existence of Restricted RPM Ranges. Radial engines are prone to internal resonance while operating within specified rpm ranges and while the engine is largely unaffected by the rpm transitioning through a so-called Restricted Range, prolonged operation within a Restricted Range will result in internal damage and possible failure.

In summary, not only must the trainee pilot understand that the engines on ZA947 are subject to many more limitations and restrictions than are commonplace on contemporary aircraft types, but that the engine controls must be manipulated carefully and sequentially to avoid damage and potential failure.

Cruise and general handling

The aircraft is pleasant to fly with powerful flight controls and highly effective elevator, rudder and aileron trimmers. The huge elevators and rudder render the aircraft particularly sensitive to both pitch and rudder inputs. Due to their size, the control loads also become quite heavy, with a relatively small deflection, so the aircraft is also equipped with powerful trim controls. The pitch trim control is a large wheel, mounted on the left side of the throttle control pedestal. The rudder and aileron trim controls are on either side of a rearward facing panel below the throttle pedestal. The trim controls enable the pilot to all but eliminate the primary flight control loads during normal flight regimes, thereby reducing physical workload and fatigue. For example, if, to maintain level flight, the pilot is required to apply a forward (nose down) force on the control column, the pitch trim wheel is turned forwards (nose down). Turning the pitch trim wheel moves a trim tab on the elevator (pitch control), which applies an aerodynamic load to the elevator in the same direction as the control force being applied by the pilot needed to maintain level flight. As the pitch trim wheel is turned, the pilot senses the reducing forward pressure required to maintain the level attitude as trim tab deflection takes effect, such that with the trim set correctly, no pressure need be applied to the control column, other than for small corrections, to maintain the desired flight path.

The rudder and aileron trim controls function in exactly the same way, enabling the pilot to cover the primary flight controls while the aircraft maintains the desired flight path. With all trimmer controls set correctly, small flight path corrections are accomplished with light pressures applied to the primary flight controls. Each trimmer control has a mechanical indicator, which displays the amount of trimmer deflection either side of the neutral position required for straight and level flight.

The large elevators, which make up well over half of the surface area of the entire tailplane, are designed to cater for the distribution of the load within the cabin and, in conjunction with the wide-ranging pitch trim, provide for a broad Centre of Gravity (CofG) envelope. All aircraft must be loaded such that the CofG envelope is retained within the certified parameters for safe aircraft control. Large jet transport type aircraft are equipped with an all-moving tailplane, which has the same effect as the pitch trim tabs on ZA947, but on a far larger scale. Next time you are boarding a jet airliner through a rear entrance door, take a look at the fuselage ahead of the tailplane leading edge. There is usually a series of markings delineating both the extent of tailplane travel and tailplane neutral position. On the Boeing E-3D Sentry, the elevator loads sensed by the pilot are trimmed out in exactly the same way as on ZA947, the only difference being an electric motor drives the flight deck pitch trim wheel as opposed to the manually turned pitch trim wheel on ZA947. In the event of a failure of the electric trim on

the E-3D Sentry, a handle is pulled out of the pitch trim wheel to turn it manually.

It must be remembered that ZA947, as a military variant of the C-47, is designed to drop personnel and cargo by parachute from the rear cargo door. As the payloads are moved aft through the cabin in readiness for the drop, the powerful tailplane elevator authority, combined with the extensive pitch trim range, enable the pilot to retain control of the aircraft as the CofG moves aft also. For the display and support activity that ZA947 undertakes with BBMF, the loadmaster ensures that any load in the cabin is retained within the cleared envelope. A Trim or Load Sheet, prepared by the loadmaster, sets out the entire load to be carried on the aircraft, including the number of crew, baggage, fuel and any freight or additional load(s). The sheet identifies not only the individual weight of each item, but also its position in relation to the

BELOW The air loadmaster wearing his dispatcher's harness, which is attached to the static line in the roof of the aircraft, stands by the rear door during take-off. *(BBMF/Crown Copyright)*

CofG. By performing a small series of simple calculations, a total weight is determined and an average 'balance' set out in diagrammatical form on the Trim Sheet for both take-off and landing.

Recovery checks

On completion of the airborne task, the RECOVERY CHECKS are performed to prepare the aircraft for landing. The normal arrival at an airfield is a visual procedure where the aircraft is positioned for the runway in use followed by a turn downwind to land or join the visual circuit if required for training purposes.

Normal visual circuits

Two types of visual circuit are flown in ZA947, the Normal Visual circuit and the Asymmetric Visual Circuit. The Normal Visual Circuit is depicted in the chart below.

The aim of both types of Visual Circuit is to

BELOW Visual circuit diagram.

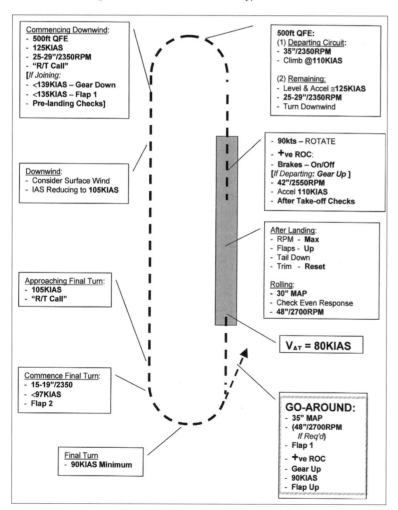

BELOW Visual circuit diagram.

Commencing Downwind:
- **500ft QFE**
- **125KIAS**
- **25-29"/2350RPM**
- **"R/T Call"**
[*If Joining:*
- **<139KIAS – Gear Down**
- **<135KIAS – Flap 1**
- **Pre-landing Checks**]

Downwind:
- Consider Surface Wind
- IAS Reducing to **105KIAS**

Approaching Final Turn:
- **105KIAS**
- **"R/T Call"**

Commence Final Turn:
- **15-19"/2350**
- **<97KIAS**
- **Flap 2**

Final Turn
- **90KIAS Minimum**

500ft QFE:
(1) Departing Circuit:
- **35"/2350RPM**
- Climb @**110KIAS**

(2) Remaining:
- Level & Accel ≅**125KIAS**
- **25-29"/2350RPM**
- Turn Downwind

- **90kts – ROTATE**
- **+ve ROC**
- Brakes – On/Off
[*If Departing: Gear Up*]
- **42"/2550RPM**
- Accel **110KIAS**
- **After Take-off Checks**

After Landing:
- RPM - **Max**
- Flaps - **Up**
- Tail Down
- Trim - **Reset**

Rolling:
- **30" MAP**
- Check Even Response
- **48"/2700RPM**

V$_{AT}$ = 80KIAS

GO-AROUND:
- **35" MAP**
- **(48"/2700RPM**
If Req'd)
- Flap **1**
- **+ve ROC**
- Gear Up
- **90KIAS**
- Flap Up

prepare the aircraft to fly an accurate approach and safe landing. It is said that a good landing requires a good approach, a statement that is particularly relevant to ZA947.

After take-off

The aircraft rotates at 90kias, as for a normal take-off to depart, but the undercarriage remains down. The wheel brakes are depressed momentarily before the pilot assumes control of the throttles and calls for the After Take-off Checks. Since ZA947 is the multi-engine training aircraft, the undercarriage is retained in the down position to avoid recurrent technical problems, as have been experienced in the past through repeated undercarriage up and down selections in the visual circuit. Nevertheless, the pilot always retains the option to raise the undercarriage for safety of flight. The engine power is reduced to 42in MAP/2,550rpm and the aircraft accelerates to 110kias for the climb to the BBMF normal circuit height of 500ft QFE (airfield atmospheric pressure reading).

Crosswind and downwind

Levelling at 500ft QFE, the aircraft is turned downwind and engine power further reduced to 28–30in MAP/2,350rpm to achieve 125kias at the beginning of the downwind leg. Abeam of the upwind runway threshold, a radio transmission is made which is typically 'Dakota 94 – downwind land' and the ATC response is typically 'Dakota 94 – surface wind two-one-zero eighteen knots.' There is no response to the ATC transmission, but the pilot must think carefully about the surface wind and analyse how it is likely to affect the aircraft on its final approach so as to anticipate rudder and aileron control inputs.

Surface-wind analysis

The RAF Coningsby runway direction is 250° so, in this scenario, the wind direction is 40° off the runway heading at a mean strength of 18kts. To enable the pilot to calculate the approximate headwind and crosswind components, a simple analogy is used that relates minutes as a proportion of an hour to degrees displacement between the surface wind and an aircraft reference point when the aircraft is 'aligned with the runway'.

The analogy starts with the assumption that anything more than 60° equates to one whole hour or, in terms of surface wind, the full strength in knots. For the crosswind component, it is in relation to degrees displacement from the *aircraft nose* when aligned with the runway. Knowing that the wind is 40° off the runway heading, 40 minutes equates to two-thirds of an hour and, therefore, two-thirds of 18kts is the estimated crosswind component, which equates to 12kts. To estimate the headwind component, the same analogy is used, but this time it is in relation to the degrees displacement from *abeam the aircraft* when aligned with the runway. There are 90° between the aircraft nose and the reference abeam. If the wind is 40° displaced from the nose, it is 50° displaced from abeam and it follows that as 50 minutes equates to five-sixths of an hour, the estimated headwind component is five-sixths of 18kts, i.e. 15kts. The pilot may now anticipate the behaviour of the aircraft on final approach.

In calm conditions, the aircraft is reluctant to slow down on the approach and the landing roll-out will require more runway to bring the aircraft under control to taxi off. The wheel brakes are not the best and it is important that the aircraft touches down at the correct speed in the right place relative to the runway threshold. However, with a headwind component, the aircraft requires more power on the approach to maintain airspeed. The aircraft will readily reduce airspeed on the approach and it is easy to be caught out and drop below the correct approach path, resulting in what the USAF pilots refer to as being a 'kinda drug-in' approach to the runway. As on take-off, gusty surface winds are a more interesting proposition. If the wind gust abates, resulting in an immediate loss of wing lift at the critical moment, the aircraft is prone to dropping suddenly with the possibility of a firm touchdown and subsequent bounce. On the other hand, if the wind strengthens suddenly, the aircraft is likely to lurch skyward again as the gust increment increases the wing lift and, typically, this will occur just as the main wheels are a few inches above the runway in the landing flare. It is not possible to predict the gust effects, but the skill is in the pilot controlling the aircraft through them.

By far the most challenging of surface winds is a gusty crosswind with little or no headwind component. There are two ways of coping with a crosswind approach to land – either the drift-corrected heading technique, also known as the 'crabbed' technique, or the 'wing down' technique, and the type of aircraft will dictate which method is the most suitable. In the 'crabbed' approach, the aircraft compensates for the crosswind component by adopting a heading slightly into wind resulting in a wings-level approach, but with the aircraft-heading offset relative to the runway approach path. The only disadvantage is that the pilot must apply a rudder correction, in the flare as the aircraft is aligned with the runway immediately prior to main wheel contact, otherwise the aircraft will jolt awkwardly as the main wheels scrub on touchdown. Most Boeing and Airbus aircraft types favour the 'crabbed' approach.

Complications

As mentioned earlier in this chapter, tail-draggers are inherently unstable in certain phases of flight and the landing roll is, arguable, the primary example. The main wheels, being ahead of the CofG, give rise to undesirable characteristics if the aircraft is not aligned perfectly with the runway on touchdown. The large rubber main wheel tyres offer relatively little rolling resistance on the runway surface when the plane of rotation is aligned perfectly with the runway centreline. The main wheel skidding action pulls the aircraft nose further askew and, crucially, the skid friction increases rapidly as the main wheel plane of rotation deviates further from the runway direction.

The drag caused by the increasing friction of the skidding of the tyres, combined with the Moment of Inertia effects of the CofG being behind the main wheels, will result in the aircraft departing out of control rapidly unless the pilot takes swift corrective action. The pilot must sense any deviation of aircraft alignment from runway direction promptly and apply decisive directional control inputs to keep the physics under control.

The aircraft is a bit like a supermarket trolley with front wheels that will not rotate properly. You have to keep your wits about you and work hard to stop the trolley from heading towards frozen foods when you want some toilet rolls. The inevitable frustration will find most people swapping their possessed trolley for another, more controllable version. Unfortunately, the same cannot be said for ZA947 and the pilot must learn to keep the aircraft under control at all times otherwise the so-called ground-loop (and an urgent need for some toilet rolls), are the inevitable consequences of allowing the natural physics to take charge.

Another undesirable characteristic of tail-draggers is that they are prone to bounce on landing, the problem being that when they start bouncing they don't want to stop. There are two main reasons for bouncing – the first is the relatively large, balloon-like main wheel tyres and the second is, once again, the main wheels being ahead of the aircraft CofG.

Pilots training to fly ZA947 are discouraged from flying a steep approach to the runway until they are fully conversant with the landing technique to minimise the risk of bouncing on landing. The aircraft requires a very shallow flare onto the runway to skim the main wheels onto the runway surface with the tail settling – for an almost three-point landing. Any significant downward vector, as the wheels make contact, will result in the aircraft stumbling skyward again. The main wheel tyres and undercarriage shock struts absorb the initial impact energy by compressing and distorting respectively, but as they do so, the moment of inertia acting through the CofG forces the rear of the fuselage downward rapidly while pivoting against the main wheels in compressed contact with the runway and forcing the aircraft into a steep, nose-up attitude. The sudden increase in nose-up attitude results in an instantaneous increase in lift from the wings just as the main wheel tyres and struts are rebounding from their distortion on landing. The combined effects of the undercarriage recoil and the increased wing lift cause the aircraft to rise away from the runway spontaneously. However, the energy imparted to the aircraft is very short-lived and as it dissipates rapidly, together with the steadily decreasing IAS, the aircraft descends sharply for another, harder contact with the runway.

The pilot must intervene swiftly, either by selecting full power and attempting to climb away or by applying back pressure on the control column to ease the rate of descent and cushion the subsequent arrival. The latter option can be hazardous, as it requires sufficient IAS to retain aircraft control without the risk of stalling and a subsequent crash-landing. If the bouncing continues unabated, the aircraft plunges to the runway with greater severity after each bounce, risking undercarriage collapse or major airframe structural damage.

Downwind to final turn

At RAF Coningsby, the normal visual circuit is to the south of runway 07/25, but to reconcile the wide-ranging training interests of the many home-based and visiting aircraft types, the BBMF visual circuit is flown to the north of runway 07/25 at 500ft QFE. The normal procedure on training sorties is for BBMF aircraft to join the BBMF visual circuit on the downwind leg from the north, remaining clear of other aircraft operating to the RAF Coningsby visual circuit to the south of the runway.

If joining downwind in ZA947, the engines are set at 28–30in MAP/2,350rpm at the beginning of the downwind leg. Following the radio transmission, the undercarriage is then lowered and the aircraft is now in the same configuration as when remaining in the visual circuit after take-off. Next, the flaps selected to ¼ down and the Pre-landing Checks completed. At this stage, the pilot will observe a steady airspeed reduction from 125kias as the aircraft approaches the final turn for the approach. As discussed previously, careful assessment of the surface wind is critical to anticipating any drift compensation on the approach and during the landing.

The final turn

At the end of the downwind leg, the throttles are reduced to about 17in MAP and a momentary level turn using 30–35° angle of bank, commenced to reduce the aircraft speed below 97kias for the selection of flap ½. With the flap ½ section confirmed, the nose is lowered to maintain 90–95kias in a descending final turn to establish on the extended runway centreline. The IAS is reduced steadily on the final approach to achieve a threshold speed of 80kias.

For a successful approach and landing

With the aircraft now in the final stages of the approach, it is important to have everything 'set'. The engines will be set at 2,350rpm with the throttles at 17–20 MAP and adjusted as necessary to achieve a progressive airspeed reduction to achieve 80kias at the runway threshold. There should be sufficient aileron into wind combined with top rudder to stop the aircraft being affected by any crosswind and, above all, to keep the fuselage aligned with the runway centreline. Although use of pitch trim is recommended on the final approach, the use of rudder trim is discouraged as it would make life very 'interesting' in the landing roll. Nevertheless, the angle of bank and amount of top rudder may be varied as necessary to allow for wind gusts, but the importance of maintaining the fuselage alignment with the runway centreline, throughout the final approach and landing, cannot be overstressed. Now comes the bit where the superior pilot uses his superior judgement to avoid a situation that will require the use of superior skill!

As the aircraft crosses the runway threshold, the aileron and rudder control inputs are adjusted continuously for crosswind compensation and to maintain fuselage alignment with the runway. A few feet above the runway, the throttles are walked back to the idle stop, at the same time as back pressure is applied progressively to the control column to enter the flare for landing and prevent the aircraft from sinking onto the runway. It is important not to close the throttles rapidly otherwise the aircraft will drop abruptly onto the runway with an inevitable bounce. To minimise the risk of bouncing, the throttle closure and application of back pressure must be carefully co-ordinated to ensure that the aircraft descent shallows with the main wheels a few inches above the runway surface as the airspeed continues to reduce. With aileron and rudder inputs retained for crosswind compensation and runway alignment, the aircraft should settle onto the runway with the upwind main wheel touching first.

There is usually a small, though reassuring, 'Erp' as the main wheel contacts the runway and once sure the aircraft is settled, the pilot calls for the flap to be raised while co-ordinating aileron and rudder inputs to maintain runway alignment, so that the other main wheel touches the runway, with a second 'Erp', followed almost immediately by the tail wheel.

With all wheels on the runway, the pilot continues to increase control column back pressure steadily to keep the tail wheel in firm contact with the runway and commences gentle braking to slow the aircraft while using rudder and differential brake for directional control. It is important not to apply the back pressure too quickly as the aircraft will become airborne again, albeit perilously close to the stall speed. The pilot must not relax, especially with a crosswind, as it is easy to induce directional oscillations and the potential for ground-looping remains until the aircraft is at a fast walking pace. Even turning off the runway can cause problems if the wind catches the tail and the aircraft weathercocks rapidly into wind.

In summary, that's all there is to it, but it is important to understand that, while the aircraft cannot be hurried in any way, both vigilance and awareness are imperative to avoiding insidious situations that will result in loss of control.

Asymmetric visual circuit

The aim of the asymmetric visual circuit is to enable the pilot to position the aircraft for a successful landing following an engine failure immediately after take-off. Other than normal fuel consumption and the dropping of parachutists, an in-flight reduction in aircraft weight is not possible in ZA947. The aircraft is not equipped with a system to jettison fuel. Consequently, all load planning is carried out to cater for an engine failure immediately after departure. The Asymmetric Visual Circuit is depicted in the diagram overleaf.

Assume engine failure after VR

The Dakota is poorly placed following an engine failure after take-off and the pilot must react quickly to ensure that the aircraft performance is not compromised unduly. The drag caused by a windmilling propeller will prevent the aircraft from achieving a positive rate of climb, so it is imperative that the failed engine is secured and the propeller feathered as expediently as possible. The so-called Immediate Actions are completed from memory to avoid delay

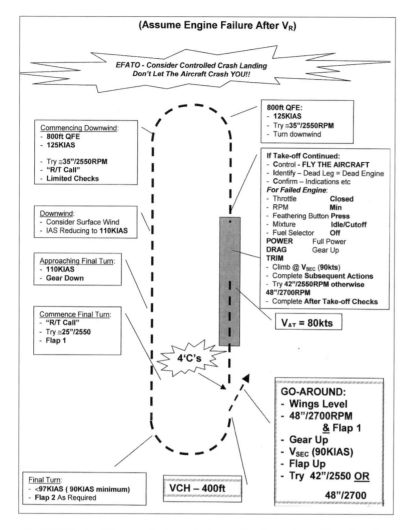

(Assume Engine Failure After V_R)

EFATO - Consider Controlled Crash Landing
Don't Let The Aircraft Crash YOU!!

Commencing Downwind:
- 800ft QFE
- 125KIAS

- Try ≅35"/2550RPM
- "R/T Call"
- Limited Checks

Downwind:
- Consider Surface Wind
- IAS Reducing to 110KIAS

Approaching Final Turn:
- 110KIAS
- Gear Down

Commence Final Turn:
- "R/T Call"
- Try ≅25/2550
- Flap 1

4'C's

Final Turn:
- <97KIAS (90KIAS minimum)
- Flap 2 As Required

VCH – 400ft

800ft QFE:
- 125KIAS
- Try ≅35"/2550RPM
- Turn downwind

If Take-off Continued:
- Control - FLY THE AIRCRAFT
- Identify – Dead Leg = Dead Engine
- Confirm – Indications etc
For Failed Engine:
- Throttle   **Closed**
- RPM   **Min**
- Feathering Button **Press**
- Mixture   **Idle/Cutoff**
- Fuel Selector   **Off**
POWER   Full Power
DRAG   Gear Up
TRIM
- Climb @ V_{SEC} (90kts)
- Complete **Subsequent Actions**
- Try 42"/2550RPM otherwise 48"/2700RPM
- Complete **After Take-off Checks**

V_{AT} = 80kts

GO-AROUND:
- Wings Level
- 48"/2700RPM
      & Flap 1
- Gear Up
- V_{SEC} (90KIAS)
- Flap Up
- Try 42"/2550 OR
                48"/2700

ABOVE Engine failure diagram.

that would jeopardise the safety of the aircraft by completing the drills with reference to the FRCs. Following completion of the checks from memory, and with the aircraft safely under control, the pilot will call for the actions to be read from the checklist to confirm that all checklist items have been completed correctly.

The 4Cs

The **4Cs** are a guide for the pilot in making a decision to commit to a landing:
The aircraft must be correctly **CONFIGURED** with . . .
A **CLEARANCE** to land on . . .
A **CLEAR** runway ahead and, above all, the pilot must be . . .
CONTENT to commit to the landing.

In the absence of any of the criteria, the only option is to go around and attempt another approach.

Post-flight

After landing, the pilot must maintain careful control of the aircraft and not attempt any other activity until the aircraft is proceeding at a fast walking pace to taxi back to the parking area. The AFTER LANDING CHECKS are completed, usually with the aircraft on the move but, if the pilot anticipates that there may be some difficulty coping with the surface wind while taxiing, the aircraft will be stopped and the parking brake applied before completing the checks. Once the checks are complete, the pilot will resume taxiing the aircraft to the parking area and, having positioned the aircraft in readiness for shutdown, the SHUTDOWN CHECKS are completed before the aircraft is handed back to the engineers.

One point worthy of note is that the right engine is the first of the two to be shut down, the reason being to check the integrity of the left engine hydraulic pump. The Dakota hydraulic system is pressurised normally by two engine-driven hydraulic pumps, one on each engine. Because the aircraft is not equipped with separate hydraulic pump pressure indicators, operation of each pump is checked independently with the respective engine running. As the right engine is started first, the right hydraulic pump is checked for pressure output before the left engine is started. On engine shutdown, the right engine is stopped first to ensure that the hydraulic system is pressurised by the left engine-driven pump. There is a mechanical hydraulic pump selector on the flight deck but, on the BBMF Dakota, the selector is retained in the BOTH position at all times in flight.

After vacating the aircraft, the crew proceed with their post-flight administration, which, for the pilots, involves completing the aircraft F700 and the Flight Authorisation Sheet. Tradition dictates that the crew then adjourn for tea and a sortie debrief.

Dakota display profile

The BBMF Dakota display is designed to present the aircraft to onlookers in its natural habitat. It is not necessary to manoeuvre the aircraft vigorously; all that is required is to take the Dakota through a series of gentle turns in relation to a fixed reference point to present both the sight and sound of the aircraft to best advantage. Usually, the Dakota display is performed as a single, stand-alone event, but occasionally, the aircraft will display with the BBMF fighters as the third and final item in a BBMF three-aircraft display. Unlike Lancaster PA474, which is better suited to the three-aircraft formation appearances, the Dakota struggles to achieve airspeeds compatible with the BBMF fighters. Consequently, if displaying with the fighters, it is not usual for the Dakota to lead fighter formations on arriving at and departing from the venue. The preference is for the Dakota to arrive from crowd-rear, to follow the fighter into the display and depart the venue as a singleton well ahead of them.

As with more routine aspects of Dakota flying operations, and aside from venue display line and orientation factors, the most important consideration for the display crew is the surface wind. The wind can have a devastating effect on the display profile and the pilot must monitor the effect of the wind throughout the display and compensate with both timing adjustments and variable turn rates to ensure that the aircraft remains anchored to the display datum.

The display datum is a visual reference point that all display participants use to position their aircraft in their respective display routines. It is usually, though not necessarily, aligned with the crowd centre, but is nonetheless something of visual significance that the crew can see easily and padlock throughout their display. At airfields it is, typically, a Day-Glo or brightly coloured marker board near the runway, otherwise, and at off-airfield venues, it is usually either a prominent or purpose-built structure. The crew must also work to a display line – in the case of the BBMF aircraft, it is the 230m line – the aim of which is to ensure that the aircraft retain a safe separation distance, in this case 230m, from the crowd at all times. As with the datum point, the venue organiser will either provide or identify visual display reference lines at specified distances from the crowd, which the display aircraft must not infringe.

Having identified both the display datum and line at the pre-flight planning and briefing stage, on arrival at the venue, the surface wind is the final variable that must be analysed carefully. Approaching the display venue, the PRE-DISPLAY CHECKS are completed and, when cleared to do so, the radio is switched to the display frequency for approval to run-in and display.

With both engines set at 2,350rpm, which other than for safety of flight is maintained throughout the display, the aircraft is positioned to run in to the venue for a fly-by along the display line at the lowest permitted, usually 100ft MSD (Minimum Separation Distance), at 150–160kias. Having identified the display datum, the stopwatch is hacked as the aircraft passes abeam and continues down the display line for a predetermined number of seconds, dependant on the surface wind. At this stage it is imperative that the pilot overcompensates, as opposed to undercompensates, for wind. If there is insufficient wind allowance, it is impossible to orientate the aircraft back to the display datum and the aircraft will be blown downwind from which it is difficult to recover. By overcompensating for wind, the pilot is able to position the aircraft back to the datum accurately.

With the datum reference time elapsed, the pilot selects 35in MAP and commences a climbing turn away from the display line to achieve approximately 500ft MSD as the aircraft passes through a reciprocal heading. Power is then reduced to maintain 120kias and the turn is continued to track back towards the datum. Tracking towards the datum, on a heading adjusted for wind, the aircraft descends to 300ft MSD towards the display line abeam the datum. Approaching the display line, a 360° turn is commenced away from the crowd aligned with the display datum. During the first half of the turn power is increased to 35in MAP and the aircraft climbs to achieve 500–700ft MSD opposite the datum and descends with reduced power during the second half of the run to achieve 300ft MSD abeam the datum.

The climb and descent during the turn is to provide a level-turn perspective for the crowd. If the aircraft maintained 300ft MSD in a level turn, from the crowd the aircraft would appear to descend (alarmingly) on the far side of the 360° turn. The manoeuvre also provides the crowd with a better opportunity for a topside perspective as the aircraft overbanks at the apex of the turn in readiness for the descending turn back to the datum. Throughout the turn, the angle of bank is adjusted to compensate for wind and maintain accurate aircraft positioning with respect to the datum.

As the aircraft passes abeam the datum, the turn is continued to achieve an outbound track of 45° from the display line in a gentle climb to 500ft MSD with reduced power, at which point flap and gear are lowered. The outbound track is maintained to achieve sufficient displacement from the display line for a descending 225° turn back towards the line for a gear-down fly-by at 100ft MSD. Established at 100ft MSD in level flight along the display line, approaching the datum the pilot applies the wheel brakes momentarily and gives a 'Gear up' instruction to the right-seat occupant. At the datum, with the gear retracting, the pilot selects 35in MAP and, at the same time, commences a climbing turn away from the datum to achieve a track of 45° from the display line as the gear retracts completely. With the gear up, the flap is raised and, having achieved sufficient displacement from the display line, the power is reduced for a final descending turn back to the display line to

achieve 150–160kias at 100ft MSD for the final fly-by along the line to either depart en route or turn downwind to land. The display duration should be exactly five minutes from hacking the stopwatch abeam the datum on the first pass, until the stopwatch is stopped as the aircraft passes abeam the datum on the final pass. A schematic diagram of the Dakota display profile is shown below.

Dakota flight-test
Before each display season, the Dakota is subjected to a pre-season flight test as set out in AP101B 4703-5M, the Dakota Flight Test Schedule. The aim of the flight test is to ascertain that the aircraft, together with all its systems and equipment, performs normally throughout all phases of flight, including those phases of flight that may be encountered in the event of one or other major system failures. The parameters to which the aircraft operates during the flight test are usually higher than those achieved during routine flight operations, the aim being to ensure that nothing malfunctions or untoward occurs as the limitations are achieved.

BELOW Display diagram.

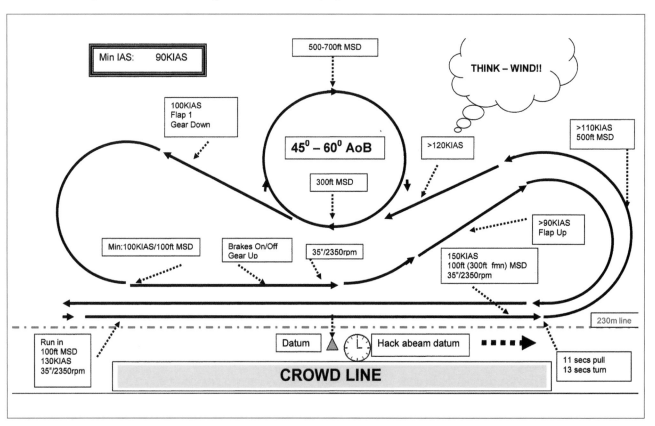

Take-off checks

Trimmers	Set
RPM levers	MAX
Throttle friction	Set
Mixture	AUTO RICH
Carburettor air	RAM and LOCK
Oil coolers	As required
Fuel:	
Contents	Checked
Cocks	Selected (normally mains)
Booster pumps	ON, pressures checked
Flaps	UP, lever NEUTRAL
Hydraulic contents	Checked
Hydraulic selector	HYD SYST (rear position)
Hydraulic pressures	Checked. Top down as required
Gills	TRAIL, then OFF
Flying controls	Full and free
Temps and pressures	Within limits
Hatches	Closed
Harnesses	Secure
Captain's brief	Complete
Transponder	ALT. Code set

Runway checks

Pitot heaters	ON
Generators	Charging (leave on BUS)
Tail wheel	Forward, locked
Landing lights	ON (only if required)

After take-off checks

Brakes	On/off

If remaining in the circuit leave the landing gear down.

If landing gear retraction required:

Latch lever	Raised
Landing gear lever	Up

Gear down line pressure zero

Landing gear lever	Normal
Flaps	Up, lever NEUTRAL
Climb power	Set
Temps and pressures	Within limits

When leaving the circuit:

Landing lights	As required
Cabin signs	As required
Altimeters	Set as required

LEFT Pitch control levers set fully forward. *(BBMF/Crown Copyright)*

LEFT Set mixture control to AUTO RICH. *(BBMF/Crown Copyright)*

LEFT Carburettor air set to RAM and LOCK. *(BBMF/Crown Copyright)*

LEFT Set oil cooler shutters as required (open or closed). *(BBMF/Crown Copyright)*

LEFT Fuel cocks select to main tanks. *(BBMF/Crown Copyright)*

Emergency equipment and drills

The Dakota holds a certain amount of emergency equipment both in the pilot's and the main compartment. It largely consists of hand-held fire extinguishers, for putting out small fires within the pilot's and main compartments, fire gloves and fire axe, which is used to break your way out of a compartment should the doors be jammed, a life raft and first-aid kits.

Escape hatches are located either side of the main compartment and one in the roof of the pilot's compartment. Also in the pilot's compartment are two knotted ropes that would enable the crew to evacuate the aircraft through the sliding windows in the pilot's compartment.

The following emergency ditching procedure is unique to the RAF; however, other DC-3s in service will have a similar set procedure in place if they are expected to fly over water during their service.

Emergency ditching procedure

Transponder   As required

If circumstances permit, ditch while power is still available. Make as near to a normal touchdown as possible, in a slight tail-down attitude, at stalling speed plus 10kts. Touchdown into wind or along the swell as dictated by sea conditions. If practicable, jettison as much cargo as possible to reduce the weight and consequent speed of impact. A low fuel state assists flotation.

Configuration

Landing gear	UP
Flaps	½ set
All external doors and hatches	Closed
Crew compartment door	Latched open

Ditching stations

Pilot	Strapped in his seat
Co-pilot	Strapped in his seat
Third crewmember	Strapped in passenger seat by rear cargo door
Passengers	Strapped in their seats

Immediate action after ditching

Third crewmember	Jettison main cabin door and launch life raft
Captain	Open hatch roof
Designated passengers	Open main cabin emergency windows

Ditching exit

All occupants	Main cabin door

Open the roof hatch and cabin side windows in case it proves difficult to open the main cabin door.

Engine fires

Should an engine catch fire on start-up, the pilot would continue to turn the engine; however, should the fire continue, and not blow itself out as often happens, the pilot would warn the crew and

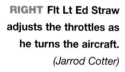

RIGHT Flt Lt Ed Straw adjusts the throttles as he turns the aircraft.
(Jarrod Cotter)

the navigator will make an emergency call to the control tower in order to summon the fire services. Meanwhile, the pilot will shut off both engines, select the fire bottle button, which would extinguish the fire. All crew will then evacuate the aircraft.

An engine fire while in the air would be dealt with in a similar manner, except that the affected engine's propeller would be feathered and the appropriate throttle, rpm lever, mixture control and fuel tank selector would be operated CLOSED and OFF. The aircraft would then make an emergency landing at the first available airfield.

Landing emergencies

Main wheel tyre failure

A normal approach would be made and the pilot would direct the aircraft to touchdown on the 'good' tyre, keeping the wing on the failed tyre side up using the aileron. He would then gently lower the failed tyre/wheel onto the runway, while aileron control remains. The rudder and brake would then be used to keep the aircraft straight.

Landing with one undercarriage leg retracted

If possible, retract the other leg and make a gear-up landing. If this is not possible use the same technique as for main tyre failure, but be prepared to use more heavy braking to keep the aircraft straight.

Wheels-up landing

In the event of neither leg descending, the pilot would carry out the same drills as for a crash-landing, which are as follows:

Transponder	As required

Use any power available to manoeuvre the aircraft to the best possible landing site. Touchdown at the lowest possible speed.

Configuration

Landing gear	UP
Flaps	Fully DOWN
Parachute door	Removed, stowed in toilet compartment
Emergency exits	Unlocked
Crew compartment door	Latched open

Crash-landing stations

Pilot	Strapped in his seat
Co-pilot	Strapped in his seat
Third crewmember	Strapped in passenger seat by rear cargo door
Passengers	Strapped in their seats

Immediately prior to touchdown

Booster pumps	OFF
Throttles	Close

After touchdown

Mixture	Both to IDLE CUT-OFF
Ignition switches	OFF
Fuel tank selectors	OFF
Fire extinguishers	Press
Battery master	OFF

NOTE: After touchdown the aircraft rolls on the retracted wheels, which protrude from the nacelles. The brakes still operate normally.

Chapter Six

The Engineer's view

In the Second World War the Dakota was renowned for being a hardwearing, durable workhorse, with a reputation for reliability. Today the aircraft is still living up to that reputation. Being easy to maintain and work on makes the DC-3 popular around the world. Chief Technician Paul Blackah has been working on the Battle of Britain Memorial Flight for 17 years where he has serviced, maintained and carried out remedial work on the Dakota. He tells us what it is like to work on this aircraft.

OPPOSITE A new engine is fitted to the engine mount on the power plant stand. *(Jim Douthwaite)*

Chief Technician Paul Blackah, MBE

When thinking about working on the Dak, the immediate thing that springs to mind is the fact that the manuals are so comprehensive and easy to read, with diagrams that are well-labelled and clear to understand. This alone makes such a difference when working with an aircraft and is a great time-saver. It also makes it easier for new members to the flight, who may not have worked on historic aircraft before, to get to grips with a task without necessarily needing to be 'babysat'. The manuals themselves were written in the 1940s and were amended up to the early 1960s.

Spares for the aircraft are still available from the United States, so it is not necessary to have parts manufactured. Prices for these spares are reasonable compared to those for other warbirds. For example, an engine overhaul costs between £25,000 and £30,000 (part exchanged), compared to approximately £100,000 for a Merlin engine; tyres are £1,200 each; and a propeller overhaul would cost about £8,500 compared to a Dowty propeller for a Spitfire or a Hurricane, which would set you back about £45,000.

Ease of access is a major plus for this aircraft. There are many quick-release panels that simply pull down, and the fuselage panels, over the wiring, can be removed by pulling out long hinge-type pins to gain access. It saves a lot of time if you don't have to remove hundreds of screws in order to remove a panel.

There have been occasions when things have not gone quite to plan and have caused a few problems, such as the time when the Dakota was in Berlin, for a Berlin Airlift celebration, and people were allowed onto the aircraft to look around. Unfortunately, one of the members of the public, while asking what certain things were in the cockpit, inadvertently pushed two buttons when asking the question and set off both engine fire extinguisher bottles. This meant that we had to remove the engine cowlings on both engines, clean all the mess from the fire bottles and then, because fire bottles are dangerous air cargo, we somehow had to get four more, complete with explosive cartridges, out to Berlin before the aircraft could fly again. This took four days to accomplish.

Other, more serious, incidents have been an engine failing during flight, a tyre bursting during landing, which caused the wheel to disintegrate and, on several occasions, she has had to be followed back by fire engines, because flames have been seen coming from the exhaust. All of these, despite sounding as if they were major problems, were relatively easy fixes and she was serviceable again within a short space of time.

Safety first

As with all aircraft, the Dakota should be treated with respect and whether static or in motion, care should be taken when moving around the aircraft. Even when the aircraft is static the propellers should be treated as 'live'. Prior to turning a propeller a crewmember in the cockpit should confirm that the magneto switches are in the OFF position or the engine could fire.

No unauthorised personnel should be close

BELOW This is the aftermath of a tyre puncture sustained during taxiing. Several holes have been made in the metal wheel rim, completely ruining the wheel. *(BBMF/ Crown Copyright)*

to the aircraft prior to starting up the engines; a propeller can easily remove a head and people not used to being around aircraft may not take the due care and attention necessary. Care should also be taken with the height of the aircraft; walking or turning into the wings, for example, can cause serious injury.

Standard safety equipment, such as safety boots, ear defenders, gloves and overalls are all necessary to protect against spillage, noise damage, slipping and injury.

Flying control locks should always be in place when the aircraft is outside. These are felt-padded wooden wedges, which stop the controls moving if it is windy, thus preventing damage to the controls and to the aircraft structure. It is important to ensure that the undercarriage locking pin, complete with warning flags, is fitted, to prevent the undercarriage collapsing on the ground should the hydraulic pressure dissipate.

Pitot/static covers should be fitted when the aircraft is not in use, which prevent any dirt or foreign bodies blocking the vents. If this were to happen it would give false indications to the aircraft's instruments.

Tools and working facilities

A comprehensive set of AF spanners and sockets is essential, as are screwdrivers, pliers and drills. More specialised equipment, such as a propeller sling and an engine-lifting beam, are also required, as well as an engine cradle to hold the engine once removed. The maintenance facility would have to be large enough to accommodate the aircraft, remembering that adequate space should be available to remove items such as the outer wings. Another consideration would be a working platform around the engine for ease of access and safety while working at height.

Although not an absolute necessity, an A-frame gantry, which is high enough to remove the propeller and the engine, would be a great asset, otherwise a crane would have to be hired.

A hydraulic testing rig, for pressure-testing and leak-testing hydraulic components, would be advantageous, or the items that require testing could be subcontracted out.

Jacking and airframe support

To jack the Dakota, two 15-tonne jacks with jacking adaptors that fit onto ball sockets, which are permanently mounted on the wing just outboard of the nacelles, are required, in addition to a tail trestle that attaches to a fixed point near the tail unit on the rear fuselage, and a pair of wing trestles.

First, the tail has to be raised with a crane. To do this, you attach the crane to the sling, which is built into the aircraft just forward of

ABOVE A comprehensive set of spanners and sockets are what is needed to work on the aircraft. *(Paul Blackah/ Crown Copyright)*

BELOW To raise the tail, the crane is connected to the lifting hoist cable at the rear fuselage. *(Paul Blackah)*

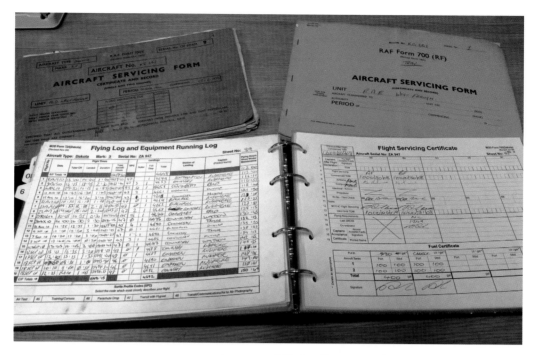

the fin. The tail is lifted and the tail trestle is positioned so the aircraft can be lowered onto it and bolted securely in place. The two jacks are positioned to allow the undercarriage to be retracted and then the aircraft is lifted off the ground. The wing trestles are put into place to steady the aircraft and enable personnel to work on the wings. Lastly, the engine staging is positioned around both the engines.

Servicing a Dakota

The servicing cycle of the BBMF Dakota is ten years between major servicing. In the years between the majors there are other less in-depth servicings. Dakotas in civilian ownership will have their own, individual, servicing schedules, which are approved by the CAA/FAA.

The servicing schedule combines a number of factors such as flying hours, landings, calendar life of components, known structural failures and taking into account the fact that the aircraft is nearly 70 years old. This schedule also includes items that are fitted to the aircraft that have their own individual 'life', regardless of whether the aircraft itself is due a service. These include items such as aircrew harnesses, which are serviced every 12 months, aircraft batteries (every three months) and main wheels (every two years).

Flight-line servicing

Due to the size of the aircraft, flight-line servicing is carried out by two people, one concentrating on the duties relating to the inside of the aircraft and the other on the outside. When the servicing has been completed it is signed for in the aircraft F700 log, on the F705 flight-servicing certificate. The pilot, who acknowledges that he accepts that the aircraft is ready for flight, then signs this form.

Before-flight servicing

Before-flight servicing is carried out within 24 hours of the aircraft's scheduled take-off time. This service is good for 24 hours afterwards, so if the take-off time is put back within that period, another service is not necessary; however, should the take-off time be even an hour over the regulated 24, then the whole service will have to be repeated. The following items are checked:

1. Internal and external panel security.
2. Flying controls for correct operation and that there are no restrictions; engine controls have the same checks.
3. Tyre pressures.
4. Windows checked and cleaned.
5. Crew harnesses and passenger harnesses.
6. Oil, hydraulic and de-icer fluids to ensure that they are at the correct level and are replenished as necessary.

7. Aircraft fuel contents are correct as required for the flight.
8. Electrical systems, such as battery voltage, navigation lights and strobe lights.

After-flight servicing

This is valid for seven days and is carried out as soon as possible after the aircraft lands, to ensure that if there are any problems they are rectified before the aircraft flies again. Problems such as minor oil leaks, loose fasteners, worn tyres and minor electrical faults can occur at any time. The checks involve:

1. Refuelling the aircraft for the next trip. This is done by two personnel: the fuel bowser driver and one of the ground crew carrying out the after-flight service. The bowser is positioned at the front of the aircraft between the engine and the fuselage. The boom is released from the bowser and the bowser is driven up to the aircraft carefully, to ensure that the boom does not hit the propeller. When in place, the hose from the boom is given to the ground crewmember who is positioned on the wing and who then also earths the fuel nozzle to the aircraft by means of an earth lead, which plugs in to an earth point on the wing. This is done to stop a build-up of static while refuelling, preventing a potential spark ignition. The bowser is then earthed to the aircraft by connecting an earth line to the undercarriage leg. The fuel caps on the wing are undone and the ground crewmember has with him two dipsticks, one for the forward tank and one for aft fuel tank. He fills the tanks, regularly stopping to check the contents by inserting the dipstick and reading off the scale, etched onto the stick, which allows him to get the correct fuel level for each tank, depending on the amount required for the aircraft. The fuel is equally distributed between the tanks; for example, if the fuel load was 400gal, there would be 100gal in each tank. When one side has been refuelled, the bowser is disconnected and the process repeated on the other side.
2. Tidying the aircrew and passenger harnesses and removing any rubbish from the main cabin.
3. Checking internally and externally for damage; looking for cracks, loose fasteners, oil leaks.
4. Cleaning the aircraft to maintain its display condition. Any oil deposited on the airframe

LEFT Overwing refuelling is carried out and the contents of the fuel tank are checked with a dipstick. *(Paul Blackah/ Crown Copyright)*

Lubricants and fuels

Fuel	Avgas/F18/100LL; total capacity – 670gal, although the majority of trips are carried out with 400gal, 100gal in each tank
Oil	OMD370/Aeroshell W120; approximately 56gal
Hydraulic fluid	OX19/H-537
De-icing fluid	AL11/S-737
Lubricating oil	OM150; used for lubricating hinges and bearings
Grease, general	XG287, used for lubricating bushes, linkages, etc.

that may have come from the engine breather pipes is removed by using fuel, as the oil can be thick and difficult to clean. The leading edges of the wings and propellers are wiped down using a detergent cleaner to remove insects that have accumulated during flight – the hotter the day, the more insects. Windscreens and windows are also cleaned using plain warm water to remove insects. Every three months the aircraft has a full wash, which involves taking the aircraft to a dedicated wash pan. All windows and doors are taped up to prevent water access and the aircraft is cleaned with pressure washers, using detergent and brushes. Care is taken not to use the detergent on the fabric flying control surfaces, as the detergent stains the fabric, which looks unsightly. The engine bay and the undercarriage bay are given particular attention due to the build-up of oil

and dirt. After washing the aircraft is given a full lubrication on the areas that have been cleaned, such as hinge points and linkages.

Turn-around servicing

This servicing is valid for 12 hours and is carried out if the aircraft is to fly again within that time period. It involves the following:
1. Refuel the aircraft as required.
2. Clean the pilot's windows.
3. Quick visual check of the exterior of the aircraft.

Engine oil levels and tyre pressures are not checked during this servicing as everything is too hot.

Primary servicing

This service is carried out at 50-flying-hour intervals and requires two technicians and takes approximately 4 to 5 hours. It consists of lubricating the main entrance doors, escape hatches and the cowl gill linkages.

Annual servicing

As the title suggests, this service is carried out on an annual basis, usually starting at the end of the display season in October, and takes approximately five to six months to complete. It includes the primary checks plus the following:
1. The internal and external structure of the aircraft is checked for corrosion, cracks, loose rivets and the condition of the aircraft's paint finish.
2. Engine cowlings are removed to allow inspection of the engines.

RIGHT All the engine cowlings are removed; here you can see both sets arranged on the floor along with other panels off the aircraft.
(Jim Douthwaite)

3. Engine oil is changed and engine filters are checked for debris.

4. Flying control cables are inspected and full 'range of movement' checks are carried out.

5. The aircraft is jacked up and the brakes are checked for correct operation and adjusted as necessary.

6. Electrical instruments and systems are checked.

7. Any 'lifed' items that are out of 'life' are replaced.

8. On completion of the servicing engine ground runs are carried out.

9. Finally, the aircraft undergoes an air test.

Engine change

One of the major tasks that could be required during an annual servicing is an engine change. This could be because the 'life' of the engine has been consumed; the engine has used its 1,250 hours as laid down in the F700 log, or during a filter check an abnormal amount of metal particles have been found, giving an indication that something in the engine has either failed, or is about to fail. The procedure for replacing an engine is as follows:

1. To remove the Hamilton Standard propeller, first take away the dome breather hole nut from the front end of the dome. Then install the dome handle, remove the lock screw from the dome-retaining nut and unscrew the nut. This nut is attached to the dome and acts as a puller when it is unscrewed. Remove the dome.

2. Remove the lock ring from the propeller-retaining nut. While two slings from a lifting gantry support the propeller, unscrew the propeller-retaining nut. The propeller is pulled off the propshaft when the propeller-retaining nut is unscrewed. Pull the propeller clear of the propshaft and carefully lower to the ground.

3. All cowlings around the engine should be removed to give accessibility to engine controls, pipes and mounting bolt connections.

4. Drain the engine oil.

5. Remove the two vent pipes from the crankcase, disconnect the oil pipe from the engine at the oil cooler.

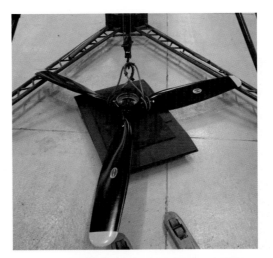

LEFT **The propeller is removed with two rope slings and lowered to the floor.** *(Jim Douthwaite)*

BELOW **All the linkages and pipe work are disconnected from the firewall.** *(Jim Douthwaite)*

ABOVE The engine is ready for lifting off the wing. *(Jim Douthwaite)*

BELOW The engine is lowered in the power plant stand. *(Jim Douthwaite)*

BELOW RIGHT The bare firewall waits for the new engine to be installed. *(Jim Douthwaite)*

6. Disconnect the inboard and outboard oil vent pipes at the first hose forward of the firewall.
7. Disconnect the main fuel pipe at the firewall.
8. Disconnect the vapour separator pipe at the firewall.
9. Disconnect the hydraulic pressure and suction pipes at the self-sealing couplings on the firewall and blank the flexible hose ends to prevent spillage.
10. Disconnect the engine fire extinguisher pipe at the firewall and disconnect the pipe from the vacuum pump, again at the firewall.
11. Disconnect the propeller governor oil pipe at the flexible pipe at the firewall and blank off the pipe.
12. Disconnect the oil in and out pipes at the firewall.

13. Disconnect the cowl flap hydraulic pipes at the firewall; this will totally free the pipes since they were disconnected from the actuating cylinder when the cowling was removed.
14. Disconnect the carburettor vapour return pipe at the firewall.
15. Disconnect the oil pressure pipe at the firewall.
16. Disconnect the fuel pressure pipe at the carburettor.
17. Disconnect the propeller anti-icing pipe at the firewall.
18. Disconnect the pressure cross-feed pipe at the firewall.
19. Disconnect the manifold pressure pipe at the firewall.
20. Disconnect the engine primer pipe at the firewall.
21. Disconnect the propeller (CSU) control cables at the turn buckles forward of the firewall, ensuring that the turn buckles are tied off, keeping tension on them, thus stopping the cables that go to the control in the cockpit from coming off their pulleys.
22. Disconnect the electrical conduit at the firewall.
23. Disconnect the carburettor air, mixture, and throttle control rods at the bell cranks on the firewall.
24. Disconnect the ignition, starter and generator conduits at the firewall disconnect plug.

25. Disconnect the engine exhaust ring from the tail pipe, attach the lifting sling to the hoist eyes on the engine.

26. Unscrew the four engine mounting bolts, but do not remove them. Approximately four and a half turns are required to free the bolts.

27. Taking the weight of the engine, remove the lower engine mount bolts.

28. Remove the two upper bolts and swing the engine forward approximately 4in to clear the exhaust stack and then pull the engine forward and lower to fit the power plant stand.

29. Items such as the generator, starter motor, carburettor and magnetos are removed from the ancillary drive at the rear of the engine. These items, if still in 'life', can be fitted to the replacement engine.

30. The engine is removed from the engine mount by removing the eight rubber-mounted bolts and is then placed into a smaller engine stand.

31. The new engine is attached to the engine mount on the power plant stand and then 'dressed' with the magnetos, starter and generator.

32. The engine is re-installed on the aircraft and the components fitted in reverse order from that of removing it. (New engine mounting bolts are attached instead of the old ones regardless of how long that engine has been in situ.)

ABOVE The engine is then removed from the engine mount.
(Jim Douthwaite)

LEFT It is then placed into an engine stand. Here you can see the old engine along with the new one ready to be fitted.
(Jim Douthwaite)

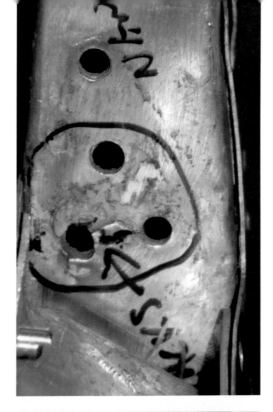

RIGHT Typical damage found to the flying control structure on ZA947 during servicing. Here you can see a crack and elongated hole. This item will need replacing. *(Dennis Lovesday)*

BELOW Repaired and replacement fittings held in place ready for riveting on the flying control. *(Dennis Lovesday)*

BOTTOM The almost completed control surface, freshly repainted. *(Dennis Lovesday)*

Minor servicing

The minor service is carried out every five years by a civilian contractor and is basically an annual servicing with the following:

1. Hydraulic components are removed and pressure tested for leaks.
2. Flight instruments are removed for servicing and tested for accuracy.

Major servicing

The major service is carried out every ten years by an outside agency that will tender for the contract. This is an in-depth service, which includes the annual, the minor and the following:

1. All the flying controls are removed and the fabric is stripped off so that the items can be inspected and a new fabric covering is applied.
2. The outer wings are removed from the centre section, so that the wing joints can be inspected for corrosion.
3. The cockpit seats are removed and the main cabin para seats are also taken out in order to enable the fuselage flooring to be lifted for inspection of the internal fuselage.
4. The undercarriage is removed and overhauled by stripping the components down, measuring for fits and tolerances, checking bore for corrosion and replacing any worn seals.
5. Electrical wiring is examined for age and condition and replaced as necessary.
6. Fuel and oil tanks are removed, flushed to remove any debris and checked for cracks using visual and non-destructive testing (NDT) techniques and then pressure tested.
7. Engine oil coolers are removed for flushing and checked for any leaks in the matrix by pressure testing.
8. All hydraulic components are removed for overhaul. They are stripped, seals are replaced, if required, the item is then reassembled and tested for correct operation.
9. During every other major service all flying control and engine control cables are replaced.
10. If the cables are not due to be replaced, they are thoroughly examined visually for corrosion and any broken strands. Checking for broken strands is done by running a soft cloth along the cable; the cloth will snag on the cable if a strand is broken. Bare hands

LEFT Main leg strut in the process of being stripped down to check for seal wear and any internal corrosion. *(Dennis Lovesday)*

CENTRE The removed unit, complete with seals, is now ready for inspection. *(Dennis Lovesday)*

BELOW LEFT Assembled inboard and outboard struts, ready for pressure testing. This checks whether there are any oil or air leaks in the assembly. *(Dennis Lovesday)*

BELOW RIGHT Aircraft main wheel, complete with brake drum. The wheel is being stripped ready for non-destructive testing, which will detect any cracks in the metal wheel. *(Dennis Lovesday)*

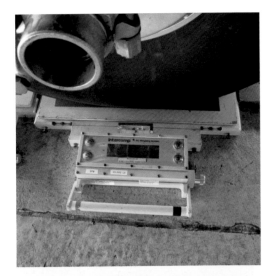

should not be used as a broken strand can
cut soft flesh and become embedded.

11. The aircraft is repainted in a new paint
    scheme either by stripping to bare metal
    (every 20 years) or by 'scuffing' the original
    paintwork.

12. The aircraft is weighed to check that the
    CofG has not altered during the major
    servicing. This can happen by removing
    redundant items or installing modified
    items, such as new radio fits; something as
    simple as a new coat of paint will affect the
    aircraft's CofG. The 'weigh' is carried out by
    having the aircraft empty of fuel, but with all
    its oil systems full. The aircraft is rolled onto
    two digital weighing pads (scales). The tail is
    then lifted with a crane and spring-balanced
    scale in order to raise the aircraft into a
    simulated flying position and the weight is
    calculated from all three scales. The aircraft
    has a set CofG range and the CofG must
    be within these limits. If it is not within range
    then ballast is added or removed, usually in
    the rear main fuselage compartment.

13. The compass swing test is carried out.

14. Engine runs are made to check the engines'
    performance and that all systems work
    correctly.

15. A full air test is carried out and, if
    successful, the aircraft is returned to the
    BBMF. If the aircraft fails the air test, for
    any reason, rectification of the problem
    is carried out and a further air test is
    completed. This process can take some
    time, should numerous problems arise, or
    can be completed on the first flight.

Non-destructive testing

Four methods of non-destructive testing
(NDT) are carried out on the Dakota: X-rays,
ultrasound, magnetic particle and dye-penetrant.
X-ray checks are conducted to examine areas
on the Dakota that cannot be easily accessed,
such as the wing internals, fin, tailplane and
engine mount. The check will usually show up
any cracking or corrosion in these areas.
Ultra-sound checks use sound waves to
detect cracking around the bore of a hole. To
do this, a probe is inserted into the hole, for
example the undercarriage mounts, where a

meter reading will indicate if there is any sign of damage around or in it.

Magnetic particle checks are used to look for cracks in ferrous metal components. The component is painted with a fluid containing iron filings and then an electric current is passed through it. When examined under ultraviolet light, any faults on the component will show up. The sample is demagnetised after the procedure.

Dye-penetrant checks are used to look for cracks in ferrous and non-ferrous components. The component is painted with a fluorescent purple dye, which is cleaned off and sprayed with developer. After approximately 40 minutes any faults will be highlighted in purple.

Keeping records

Keeping an accurate servicing and flying record is essential for any owner/operator. Civilian aircraft are governed by the CAA, which operates a system similar to the RAF. The RAF uses a Form 700 logbook, which contains various forms relating to the aircraft's flying hours (F724), faults (F707A) and the aircraft's flight servicing certificate (F705).

The F724 and the F707As are kept for the time that the aircraft is in active service. Major components, such as engines, propellers and magnetos, have their own log cards, which stay with the particular component. These cards show the date of the fitment/removal of the item, the flying hours of the item, the date it is due for overhaul and details of any repairs or modifications that have been carried out.

Choosing a new colour scheme

After each major service the BBMF Dakota colour scheme is changed to represent a particular squadron that saw active service during the Second World War. After research, up to three suggestions are put forward by members of the flight and a final decision on the squadron, and on a specific aircraft in that squadron, is made by the Officer Commanding BBMF. Photographs and surviving crewmembers' recollections ensure that the finished markings and codes are accurate. The BBMF is always grateful for the general public's help in achieving this.

Manuals

To assist with servicing the Dakota, a set of manuals is essential. These cover all aspects of servicing the Dakota and are known as the AN-01-40 and the T.O.1C-47A series of books. These books cover maintenance instructions, illustrated parts catalogues, scheduled maintenance, basic weight checks, flight operating instructions and structural repair instructions. Other manuals include engine overhauls and parts, and undercarriage/ brakes overhauls and spares.

Aircraft part numbers

The Dakota has its own unique part number system consisting of a seven-digit number, which can start either 111, 211 or 511. If an item is used on the left- and the right-hand sides of the aircraft, the part number for the right-hand number finishes with '–1'. For example, 5117072 is a cover left-hand and 5117072–1 is a cover right-hand.

Nuts, bolts and washers are prefixed with the letters AN, so the first digit of the bolt denotes the diameter in sixteenths of an inch. For example, an AN3 bolt is three-sixteenths in diameter and an AN7 bolt is seven-sixteenths in diameter. The dash number works by denoting the whole length, including the thread, in eighths of an inch in a way that only Americans could have invented. The easiest way to explain this is by example: an AN3–12A denotes a bolt, which is not twelve-eighths of an inch long, but one whole inch and two eighths.

Undercarriage and brake components are supplied by Bendix and have their own unique five-digit part numbers starting with either 5 or 6. When the Dakota was operated by the RAF, the prefix 126JQ was put in front of the part number, making the part number both a section and reference number.

LEFT Data plates are on all the aircraft's large components; this one is from the wing tip. It shows the modification plate, stamped with any modifications that are embodied in the component. The small plate gives a serial number and the date of manufacture. *(Paul Blackah/ Crown Copyright)*

Repainting the aircraft

An aircraft should always be stripped down to bare metal before repainting to prevent the build-up of paint layers, which can affect the weight and handling. The Dakota is a particularly large aircraft and it is not always possible to find a company that has the facilities to both service and strip her at the same location. This means that after servicing, or when a new paint scheme is planned, the BBMF aircraft has to go to a specialist aircraft paint company.

After major servicing in 2010–11, the BBMF Dakota was flown to Manchester for repainting. On arrival, all areas such as windows and fabric flying control surfaces were masked off and the aircraft was chemically stripped of paint. Following an inspection of the aircraft skin, a layer of primer was applied to allow the topcoat to bond to the aircraft. Once dry,

and within a 24-hour period, the topcoat was applied. In the case of the BBMF Dakota, the topcoat is now dark green for the upper surfaces and grey on the under surfaces.

To achieve the effect of the camouflage line, a technique called 'wet on wet' was used; the camouflage line was sprayed in either green or grey, by hand, and immediately the other colour was sprayed up to the previous line, giving a feathered effect. Once dry, the roundels were measured out and masked off. One colour was applied, allowed to dry, re-masked off and the next colour applied, and so on for a four-coloured roundel. Invasion stripes were added in a similar manner, although in wartime they were painted on simply by using a wide brush with minimum care.

Finally, the aircraft windows were de-masked,

BELOW The aircraft is paint-stripped and all the flying control surfaces are masked off ready for the primer to be applied.
(Ian Hopwell)

flying controls uncovered and the stencil details, such as emergency hatches, oil and fuel fillers, earthing points and warning signs were applied, as was the nose art detail. A two-pack acrylic paint and hardener was used on the metalwork and cellulose on the fabric.

The cost of a total strip and re-spray on the Dakota is £60,000–£80,000.

RIGHT Looking down on the nose of the aircraft in its stripped condition. The windscreen wipers and the windows are masked off. *(Ian Hopwell)*

BELOW ZA947 showing off her invasion stripes. During the D-Day period the stripes would have been applied with a bass broom and would not have looked as neat and uniform as in the picture. *(Paul Blackah/Crown Copyright)*

Appendices

```
━━━━━━━━━(●)━━━━━━━━━━━━━━━━━━━━━━━
```

Appendix 1

RAF Dakota squadrons

Appendix 2

Restored Dakotas

Appendix 3

Glossary and abbreviations

Appendix 4

Useful addresses

OPPOSITE Wings Ventures' DC-3 D-Day veteran, N1944A, flies over the English countryside. *(Richard Paver)*

Appendix 1

RAF Dakota squadrons

The RAF Dakota squadrons listed below were involved in a variety of different operations during the Second World War. Space is not available to credit every action, but we hope the following résumé will provide a brief glimpse of the roles they played.

24 Squadron
Dakotas were delivered to 24 Squadron at Hendon in March 1943. The squadron was initially in Ferry Command, which eventually became Transport Command. On D-Day, 6 June 1944, the squadron carried out resupply flights to front-line troops on the Normandy bridgehead, evacuating the wounded on the return trips. Dignitaries, politicians, senior members of the armed forces and, most notably, Prime Minister Churchill and his wife, were ferried by 24 Squadron's Dakotas.

31 Squadron
DC-2s were operated by 31 Squadron in India between April 1941 and April 1943, with the squadron also acquiring DC-3 Dakotas in April 1942. They worked the routes into Burma, flying between Calcutta and Rangoon to drop supplies to the 14th Army.

48 Squadron
The squadron took delivery of its first Dakota shortly after moving to RAF Down Ampney, Gloucestershire, in February 1944. During the night of 5/6 June 1944, 30 Dakotas from 48 Squadron joined a formation of 72 aircraft to fly airborne troops across the Channel. Between 17 and 24 September the squadron towed Horsa gliders carrying troops of the British 1st Airborne Division and flew resupply sorties in Operation 'Market Garden'. The squadron was disbanded in India on 16 January 1946 and re-formed on 15 February at Kallang,

Singapore, renumbered 215 Squadron. It used Dakotas until 1951.

52 Squadron
Re-formed at Dum Dum, India, in July 1944, from C and D Flights of 353 Squadron, 52 Squadron worked 'The Hump' into Burma. The squadron continued to operate in India and Burma until transferred to Kuala Lumpur, Malaya, in November 1948. Dakotas remained in service with the squadron until 1951 when they were replaced by Vickers Valettas.

76 Squadron
The squadron flew Hampdens and Halifaxes in Bomber Command throughout the Second World War until May 1945 when it was transferred to Transport Command. Re-equipped with Dakotas, 76 Squadron moved to RAF Broadwell, Oxfordshire, in August 1945. The following month it went to India, where it was disbanded a year later.

77 Squadron
The squadron flew Whitleys and Halifaxes in Bomber Command throughout the Second World War until July 1945 when it was re-equipped with Dakotas and transferred to Transport Command. It moved to RAF Broadwell, Oxfordshire, in August, before going to India the following month, where it was disbanded in November 1946.

78 Squadron
The squadron flew Whitleys and Halifaxes in Bomber Command throughout the Second World War until June 1945 when it was re-equipped with Dakotas and transferred to Transport Command. It moved to the Middle East in September where its Dakotas served in a transport capacity until replaced by the Valetta in 1950.

96 Squadron

Serving the majority of the Second World War as a fighter squadron, 96 Squadron was disbanded in December 1944 and re-formed in Transport Command, operating the Halifax C Mk III, followed by the Dakota from April 1945. The squadron then left for the Far East where it played a significant role in the invasion and liberation of Burma.

117 Squadron

The squadron was re-formed at Khartoum in April 1941 and began conversion to the Dakota in October 1941. Initially, it operated as support and transport for the troops in North Africa, before moving to Sicily, and then on to India where it was trained to drop paratroops. The squadron saw action in some of the worse climatic conditions imaginable, and its

contribution over four and a half years was incalculable.

194 Squadron

Formed in October 1942 at Lahore, Punjab, 194 Squadron began conversion to the Dakota in May 1943. It took on the Indian National Airways Mail Service as well as acting as a supplies carrier across the Far East, and as a means of dropping paratroops. The squadron was disbanded in February 1946, having taken part in significant campaigns across the Far East.

233 Squadron

233 Squadron spent the majority of the Second World War based in Gibraltar where it operated Lockheed Hudsons on anti-submarine duties. It returned to the UK in March 1944 and converted to Dakotas in the transport role.

ABOVE ZA947 in 77 Squadron markings as YS-H. *(BBMF)*

BELOW ZA947 in her 2011 paint scheme over the Lincolnshire countryside. *(Jarrod Cotter)*

On D-Day the squadron towed gliders and dropped paratroops, and carried out numerous resupply sorties. In March 1945 it took part in Operation 'Varsity'. The squadron went on to fly general transport sorties in South-East Asia before it was absorbed by 215 Squadron in December 1945.

267 Squadron
Re-formed at Heliopolis on 19 August 1940 and used for air transport duties in Egypt, 267 Squadron operated various aircraft, including Dakotas, across Europe and the Balkans. It was transferred to India in February 1945 where it flew Dakotas until disbanded in July 1946.

435 (Canadian) Squadron
The squadron was formed with Dakotas at Gujarat, India, in December 1944, and was instrumental in supporting the 14th Army in the Far East. After returning to Down Ampney, Gloucestershire, in August 1945, it flew on air transport duties with two other Canadian squadrons, 436 and 437, until March 1946 when the squadron was disbanded.

512 Squadron
Formed at Hendon, London, on 18 June 1943, 512 Squadron operated Dakotas on supply

routes from the UK to Gibraltar and North Africa. On 14 February 1944 the squadron moved to RAF Broadwell, Oxfordshire, where it was trained to tow gliders and drop paratroops. Its Dakotas carried elements of the 3rd Parachute Brigade to Normandy on the eve of D-Day and saw action at Arnhem during September 1944 in Operation 'Market Garden' when it towed Horsa gliders containing men and equipment of the 1st Borders. The squadron was disbanded in Italy in March 1946.

575 Squadron
Formed at Hendon, London, on 1 February 1944, 575 Squadron moved shortly afterwards to RAF Broadwell, Oxfordshire, for further training. On the eve of D-Day, 5 June 1944, the squadron dropped elements of the 3rd Parachute Brigade into Normandy in Operation 'Tonga', and on the evening of the 6th the squadron towed Horsa gliders to Normandy. In September 1944 it was involved in 'Market Garden' at Arnhem when it carried troops of the 1st Borders and flew resupply operations. After the Second World War 575 Squadron operated as a transport unit between the UK and India, and subsequently between Italy, Austria, Romania, Greece and Bulgaria. The squadron was disbanded at Kabrit, Egypt, in August 1946.

BELOW Overhead shot of ZA947 in 267 Squadron markings. *(BBMF/ Crown Copyright)*

Appendix 2

Restored Dakotas

Across the world Dakotas are still working in practical roles, not only as display aircraft. Below is a selection of the aircraft that are operating in the UK at the time of publication.

Dakota DC-3 N147DC

Owned and operated by Aces High Ltd, N147DC is available for hire to appear in television and film productions, and advertisements. It has been seen in *Band of Brothers*, *The Dirty Dozen* and *Tenko*. This aircraft can also be seen at various air shows during the year (see Chapter 4 for a full history).

Dakota Mk III N473DC

Currently based at the Aviation Heritage Centre, Lincolnshire, N473DC is available to book for a taxi run. The aircraft takes part in commemorative events and air shows (see Chapter 1 for a full history).

Dakota C-47 G-AMPY

The C-47 destined to become G-AMPY was originally allocated USAAF serial 43-49308, but was delivered in November 1944 as KK116 and flown in India by the Canadian crews of 435 Squadron (RCAF), dropping supplies to the 14th Army on the Burma front. When retired from military service it was stored in the UK at 12MU Kirkbride, until sold and allocated civilian registration. It became G-AMPY in 1952 when it started a new life in the ownership of a succession of independent airlines.

Initially operated by Starways, it was leased to Air Jordan as JY-ABE and then sold to

BELOW N473DC in the air, undercarriage retracting. *(Jarrod Cotter)*

ABOVE G-AMPY (KK116) retracts its undercarriage on climb-out from Coventry on 15 July 2008, on the occasion of the last fare-paying passenger flight conducted by an Air Atlantique DC-3.
(Tim Badham)

RIGHT G-AMPY with a light covering of snow at Coventry in January 2010. Beneath the fuselage can be seen a wind-driven pump used for dispersant fluid for the spray modification. The aircraft has dispersant tanks inside, and a green-painted spray boom outside, at the rear. G-AMPY is on stand-by for pollution control duties.
(Tim Badham)

Icelandair as TF-FIO. Returning to the UK register in 1965, it was leased to Irefly and then in 1968 was repainted as N15751 for US-based company Aviation Enterprises, but the aeroplane never left the UK.

By 1970 the aircraft was restored as G-AMPY and was operating with Intra Airways of Jersey, whom it served well for ten years. A planned move to Ireland for Clyden Airways as EL-BKJ fell through and it moved on again.

Ownership of the airframe by Air Atlantique began in 1982 and G-AMPY has remained with Air Atlantique or its associated companies ever since, although the aircraft has occasionally been leased out.

For a short spell in 1986 G-AMPY took on the guise of a Northwest Airlines example for an anniversary of the US airline. It was painted in the airline's 1940s colours for publicity photographs and at a display it flew in formation with one of the airline's Jumbos.

During the aircraft's years based at Coventry, many hundreds of people enjoyed flights in this stalwart, including those privileged to be aboard when, in 2008, it performed the last fare-paying passenger flight by an Air Atlantique Dakota. Today the aeroplane is once again configured as an aerial sprayer ready for anti-pollutant work although, as a tribute to its military origins, G-AMPY sports RAF post-war type colours and carries its wartime serial KK116.

Dakota C-47 G-AMRA c/n 15290/26735

This C-47 was allocated USAAF serial 43-49474, but was delivered in December 1944 to the RAF as KK151. After being ferried out to Australia and allocated to 243 Squadron based at Camden, NSW, it was flown on communications support flights to British Pacific Fleet bases. When that unit disbanded the aircraft returned to the UK. By mid-1946 KK151 was with Abingdon-based 525 Squadron (later 238 Squadron), flying mail and newspapers to British bases on the Continent. The aircraft next served with Oakington-based 46 Squadron, which was involved in the Berlin Airlift operating from Wunstorf, Fassberg and Lubeck. By September 1949 KK151 was withdrawn from military use and ferried to 12 MU at Kirkbride for storage.

The aircraft was allocated civil registration G-AMRA in 1952 and (like its sister-ship G-AMPY) was obtained for use by Starways. It was not long before G-AMRA was back in military colours. After sale to Airwork it was used on trooping charters and so, for a while, carried military serial CE820, but then reverted back to its civil identity. It has remained on the UK register throughout its civil career even though it has had numerous operators, including Aden Oil Refining, BUA, Morton Air Services, British Island Airways, Macedonian Aviation, Humber Airways, Intra Airways and Eastern Airways. It was obtained by Air Atlantique in 1981.

G-AMRA has operated as a passenger flyer and freighter since being based at Coventry. It has also featured in a number of films and TV shows. Before joining Air Atlantique it had appeared in the 1978 production *Atom Spies*, painted as CCCP-7245, a Soviet airliner. Then in 1991 it carried Empire Airways colours for an episode of *Poirot*. The aircraft is also believed to have appeared as one of the C-47s in *Band of Brothers*. This airframe must have a liking for the spotlight, as it is still occasionally called on to provide period dressing for TV or films.

BELOW G-AMRA displays its agility during an air-show performance at Sywell, in August 2010. *(Tim Badham)*

Dakota C-47 G-ANAF c/n 16688/33436

This C-47 was allocated USAAF serial 44-77104, but was flown to the UK in mid-1945 as KP220 initially for use by 435 (RCAF) Squadron, which operated into Europe. In March 1946 it transferred to 46 Group Communications Flight and a year later moved to communications duties with 24 Squadron at Bassingbourn. In 1950 its military service ended with a period of storage at 22MU at Silloth until sold for civilian use.

The aircraft's civil career began when it was registered as G-ANAF for use by BKS Air Transport, which operated it for five years from 1953 to 1958, before being sold to Hunting for aerial survey work. In 1973 it was moved to Duxford and preserved with the East Anglian Aviation Society, but in a surprise development in 1977, it was reactivated, passing to West Country Aircraft Servicing of Exeter for refurbishment, then registered to Air Atlantique in October that same year. It was used, largely, for mail flights and was similarly used by Air Luton to whom it transferred in 1985, passing a year later to Top Flight. The new owner only

operated it for a while before the aircraft endured a period of storage back at Exeter.

From May 1987 the aeroplane was on the US register for a year as N170GP, but it was retained in Europe. During this time it sported a smart blue and white paint scheme for a cameo role in the film *Raiders of the Lost Ark*. The aircraft was then to see a second period with Air Atlantique from June 1988 to the present day. After an initial spell as a pollution control sprayer, it was adapted for use as a radar equipment test platform. This conversion was quite dramatic as it included the fabrication and fitment of an under-fuselage radome (from a B-N Defender), the installation of a fuselage-mounted jet APU and internal equipment racks. Shortly after this it spent a few weeks in water-soluble olive drab paint and RAF colours for use in a film about the Second World War, but soon reverted to Air Atlantique's house colours of white, black and green stripes. When appropriate it has sported various clients' brandings, which have included those of RACAL and Thales; in 2007 it was repainted in a stunning red and black scheme. It remains a very active aeroplane.

BELOW G-ANAF doing a night ground run in December 2009 at Coventry. The radome, used to house radar scanners, is evident beneath the fuselage.
(Tim Badham)

Appendix 3

Glossary and Abbreviations

ABC automatic boost control
ASI airspeed indicator
ATC Air Traffic Control
AUW all-up weight
BBMF Battle of Britain Memorial Flight
boost The pressure of the fuel–air mixture in the induction manifold of the engine, measured in pounds per square inch on the boost gauge in the cockpit; this indicates the power level delivered by the engine.
CHT cylinder head temperature
CofG Centre of Gravity envelope
CSU constant speed unit – this unit, fitted to the engine, automatically controls the engine rpm and the propeller pitch mechanism so that they operate at high efficiency over a wide range of flight conditions. The pilot sets the required engine rpm using his control lever, and a governor automatically controls the pitch of the propeller blades. Thus the aircraft flies at the most efficient value for the engine power selected and the speed of the aircraft.
DERA Defence Research Agency, the successor to the Royal Aircraft Establishment.

DME distance measuring equipment – gives the distance to the transmitter.
F700 Form 700 aircraft logbook
GPS Global Positioning System
IAS indicated airspeed
IFR Instrument Flight Rules
ILS Instrument Landing System
MRCA multi-role combat aircraft
MSD minimum separation distance
NDT non-destructive testing
NOTAMS Notice to airmen – an advisory notice giving information on the establishment, condition or change in any aeronautical facility, service, procedure or hazard.
Pax passengers
POB people on board
QFE airfield atmospheric pressure reading
RAE Royal Aircraft Establishment
RCAF Royal Canadian Air Force
rpm revolutions per minute
trim tab A small control surface that, when set, enables the aircraft to fly straight and level without any forces on the control column.
USAAD United States Army Air Defense
USAAF United States Army Air Force
VFR Visual Flight Rules

Appendix 4

Useful addresses

Aces High
Dunsfold Aerodrome
Stovolds Hill, Cranleigh, Surrey GU6 8TB
Tel: 01483 200057
www.aceshigh-aviation.com

Aerospace Support Ltd
21B Warpsgrove Lane
Chalgrove, Oxford, Oxfordshire OX44 7RW
Tel: 01865 400106
www.aerospaceintl.com
Dakota spares.

Air Atlantique Classic Flight
Dakota House
Coventry Airport West
Coventry, Warwickshire CV8 3AZ
Tel: 024 7688 2616
www.classicflight.com
Operates the Dakota and other aircraft as a historic flight.

ARCo (The Aircraft Restoration Company)
Building 425,
Duxford Airfield, Duxford
Cambridgeshire CB2 4QR
Tel: 01223 835313
www.arc-duxford.co.uk
Bay servicing of hydraulic components and structural repairs.

Basler Turbo Conversions LLC
255 West 35th Avenue
PO Box 2305
Oshkosh
Wisconsin 54903-2305
USA
Tel: (920) 236-7820
www.baslerturbo.com
Fits turbo-props and modern systems to Dakotas.

CFS Aeroproducts Ltd
The Alvis Works
Bubbenhall Road, Baginton, Coventry
Warwickshire CV8 3BB
Tel: 024 7630 5873
www.cfsaeroproducts.co.uk
Engine repair and propeller overhaul.

Delta Jets Ltd
Hangar H1, Cotswold Airport
Gloucestershire GL7 6BA
Tel: 01285 771494
www.deltajets.com

Eastern Airways
Humberside International Airport
Kirmington, North Lincolnshire DN39 6YH
Tel: 01652 688456
www.humbersideairport.com
Aircraft servicing.

Lincolnshire Aviation Heritage Centre
East Kirkby Airfield
East Kirkby, nr Spilsby, Lincolnshire PE23 4DE
Tel: 01790 763207
www.lincsaviation.co.uk
The museum operates a taxiing Lancaster and a flying Dakota, which is also available for taxi runs.

Vintage Fabrics
Mitchell Hangar, Audley End Airfield
Saffron Walden, Essex CB11 4LH
Tel: 01799 510756
www.vintagefabrics.co.uk
Fabric for the flying control surfaces.

Watts Aviation Services Ltd
Church Road, Lydney
Gloucestershire GL15 5EN
Tel: 01594 847290
www.wattsaviation.co.uk

Index